Alicyclic Chemistry

Martin Grossel

Department of Chemistry, University of Southampton and Christ Church, Oxford

Series sponsor: **ZENECA**

ZENECA is a major international company active in four main areas of business: Pharmaceuticals, Agrochemicals and Seeds, Specialty Chemicals, and Biological Products.

ZENECA's skill and innovative ideas in organic chemistry and bioscience create products and services which improve the world's health, nutrition, environment, and quality of life.

ZENECA is committed to the support of education in chemistry and chemical engineering.

OXFORD NEW YORK TOKYO
OXFORD UNIVERSITY PRESS
1997

Oxford University Press, Great Clarendon Street, Oxford OX2 6DP

Oxford New York
Athens Auckland Bangkok Bogota Bombay
Buenos Aires Calcutta Cape Town Dar es Salaam
Delhi Florence Hong Kong Istanbul Karachi
Kuala Lumpur Madras Madrid Melbourne
Mexico City Nairobi Paris Singapore
Taipei Tokyo Toronto Warsaw

and associated companies in
Berlin Ibadan

Oxford is a trade mark of Oxford University Press

Published in the United States
by Oxford University Press Inc., New York

A catalogue record for this book is available from the British Library

Library of Congress Cataloging-in-Publication Data
(Data applied for)
ISBN 0 19 850104 8

Typeset by the author

Printed in Great Britain by Bath Press Ltd., Bath

Series Editor's Foreword

The ability to understand chemistry in three dimensions is of fundamental importance. *Alicyclic Chemistry* provides an ideal template for gaining an understanding of 3D structure, dynamics, and reactivity of organic architectures. Alicyclic chemistry is an important topic in its own right since many natural products and modern pharmaceuticals are based on alicyclic ring systems.

Oxford Chemistry Primers have been designed to provide concise introductions relevant to all students of chemistry and contain only the essential material that would usually be covered in an 8–10 lecture course. In this Primer, Martin Grossel has produced an excellent pedagogical account of this fundamental area, which is clear and easy to read. This Primer will be of interest to apprentice and master chemist alike.

Stephen G. Davies
The Dyson Perrins Laboratory
University of Oxford

Preface

I have enjoyed alicyclic chemistry since my days as a Ph.D. student in the University of London. This primer has evolved from lecture courses and tutorials given in Oxford and elsewhere. I have many people to thank for their help and encouragement, principally Prof. John Perkins and Dr Gordon Whitham. Both of them have, in their different ways, profoundly influenced my career and my chemical interests and I am particularly grateful to John who provided valuable advice about this text (though the contents are entirely my responsibility!). Dr Jonathan Essex and Dr Ian Stevens at Southampton also provided valuable comments. In addition I should like to thank a number of others, particularly Robert Parker and other members from my research group at Southampton; and Katherine Crapnell, Andrew Bond, Fred Boardman, and David Jones, and the many other former and recent Christ Church Chemists who have made helpful comments, provided encouragement, and nagged me to finish the primer so that they could write their essays!

I am also very grateful to my family for their support.

Finally I should very much like to thank Prof. Steve Davies for his patience!

Southampton M. C. G.
July 1997

Contents

1. Introduction

1.1 What is alicyclic chemistry?

Alicyclic chemistry is the chemistry of organic compounds containing one or more non-benzenoid rings. The scope of the current text will focus on saturated and partially unsaturated carbocyclic ring structures but brief mention will also be made of saturated heterocyclic systems where appropriate. Whilst the major emphasis will be placed on monocyclic derivatives, some bi- and polycyclic systems will also be considered.

Alicyclic structures play a major rôle in organic chemistry. Many important natural products exploit saturated or partially unsaturated rings, either as the main structural framework *e.g.* in terpenoid derivatives such as limonene (in citrus fruits) and menthol (in mint), or have rings fused together to form more complex structures as seen in the backbone skeleton of steroids like progesterone.

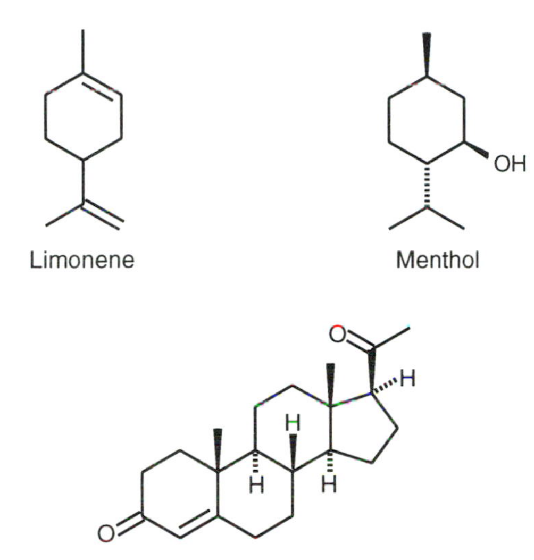

Limonene

Menthol

Progesterone - a female hormone

Small cyclic terpene derivatives provide scents in the plant kingdom. (+)-Carvone smells of spearmint, whereas its enantiomer has the fragrance of caraway seed.

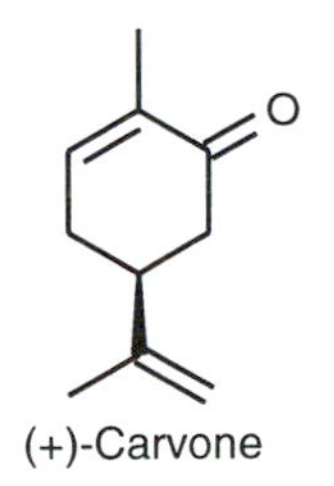

(+)-Carvone

Large carbocyclic ring derivatives are found in pheromones such as muscone and have thus been important to the perfumery industry for many years.

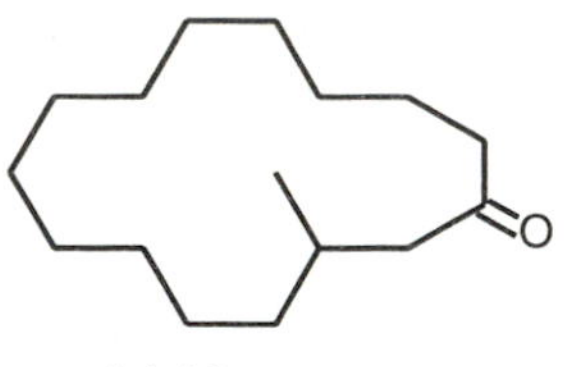

(±)-Muscone

More recently there has been great interest from many areas of chemistry in large macrocyclic *(i.e.* large ring) structures containing several heteroatoms because of their abilities to selectively complex metal and other ions; these are the crown ethers discovered by Charles Pedersen in 1962 for which he received a Nobel Prize in 1987.

Crown ethers are large macrocyclic rings which adopt crown-like geometries when they complex metal ions:

K^+ complexed with 18-crown-6

12-Crown-4

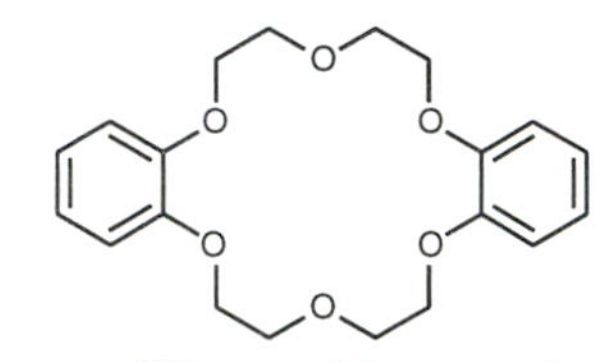

Dibenzo-18-crown-6

Ring compounds demonstrate a number of special properties which make them particularly valuable and rewarding to study. These include unusual reactivity, a variety of molecular shapes which in some cases are rigid whilst in others are very flexible, and stereochemical control of behaviour which can be manipulated by the presence of other functional groups. In addition, the synthesis of ring compounds provides a number of unusual challenges which require the application of special techniques.

1.2 Nomenclature

The names applied to simple ring structures are generally derived directly from those of their acyclic analogues with "*cyclo*" added as prefix. For example, the 6-membered ring alcohol shown below is called cyclohexanol but when a carboxylic acid group is attached to a carbocyclic ring the compound is referred to as the cycloalkane carboxylic acid *not* the cycloalkanoic acid.

The names of alicyclic compounds are directly analogous to those of their open chain analogues:

e.g. the saturated four-membered ring is called cyclobutane

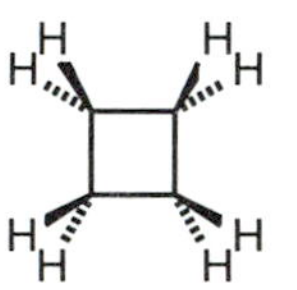

and the five-membered cyclic ether derived from hydrogenation of furan is known as tetrahydrofuran (or THF).

O
H_2C CH_2
H_2C $-CH_2$

Cyclohexanol

Cyclohexane carboxylic acid

Substituents are numbered according to their position around the ring with the usual conventions of using the lowest numbers and giving the lowest number to the main functional group *e.g.* -OH. The description of isomers will be discussed in chapter 2.

Alicyclic compounds are not limited to one ring. Polycyclic structures of various types can also be constructed. These can be divided into three general groups: spiro, fused and bridged systems.

In a *spiro* structure two rings are linked through one atom:

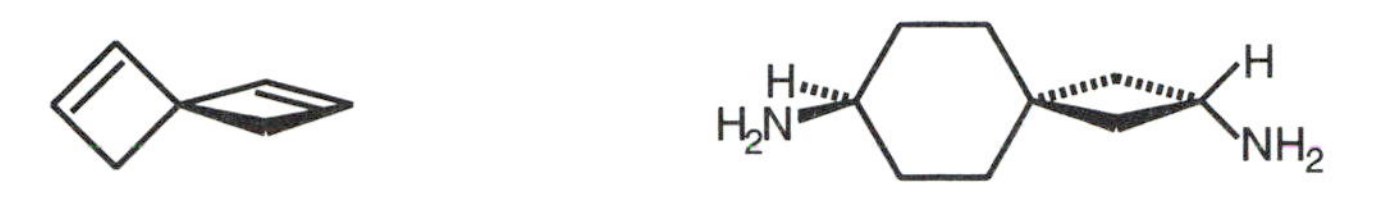

This forces the two linked rings to be held in planes perpendicular to each other and gives rise to the possibility of optical activity in both the structures shown above.

A *fused* structure contains two rings linked through a common bond. For example, decalin, which is obtained from the hydrogenation of naphthalene, consists of two fused cyclohexane rings. Such structures are not planar and in many cases can adopt several geometries showing markedly different properties. Indeed whilst *trans*-decalin is rigid, the *cis* isomer is conformationally flexible.

Decalin consists of two fused cyclohexane rings which can be linked in a *cis* or *trans* manner.

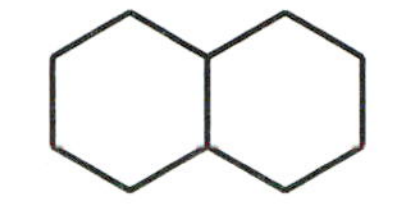

These two isomers not only have dramatically different overall shapes but the *cis*-isomer is flexible whereas the *trans*-fused structure is rigid. This results in marked differences in the behaviour of many of their derivatives.

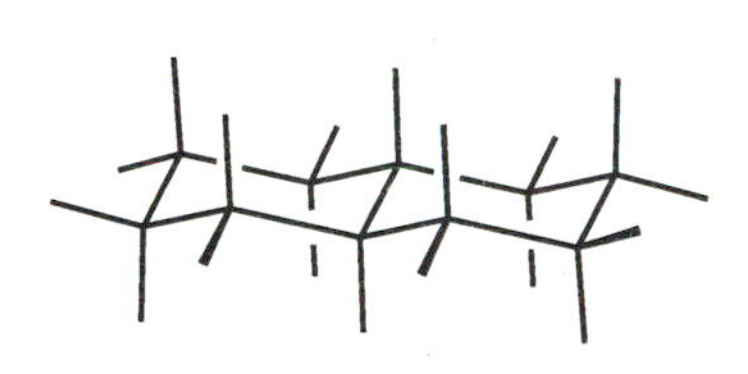

trans-Decalin

cis-Decalin

Bridged ring systems consist of several rings joined through two common atoms which are not necessarily adjacent to each other. Some examples are shown below:

Norbornane

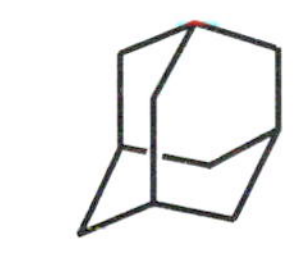

Adamantane

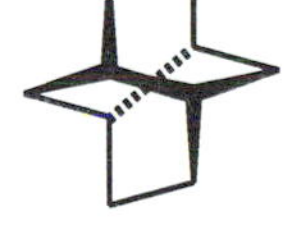

Twistane

Compounds of this type have proved a rich source of interest to organic chemists. For example, they can provide rigid frameworks which allow the investigation of the effect of geometry on reaction pathways and the interactions between different functional groups. In addition, such structures are often significantly distorted or "strained" as a result of which they frequently have interesting and unusual chemical properties.

The naming of bi- and polycyclic structures clearly adds another level of complexity to the problem of nomenclature and this will be discussed briefly in Chapter 5.

1.3 Notes and comments

Stereochemical conventions:

A wedge is used to indicate a bond coming out of the plane of the page;

a hatched line implies a bond below the page;

and wiggly line is used to when the stereochemistry is undefined.

The usual conventions have been used to provide an indication of stereochemistry *i.e.* a solid wedge implies a bond coming out of the page and a hatched line leads to a group behind the page. A squiggly line implies undefined stereochemistry.

At times in this text it is necessary to consider orbital interactions between various groups and molecular fragments. Orbital interactions are discussed using simple Frontier Molecular Orbital (FMO) theory, a basic knowledge of which is assumed.

The reader is strongly encouraged to have molecular models to hand while working though this text since the spatial arrangements of groups are an important factor in determining the course of reactions of alicyclic systems.

2. The conformational analysis of alicyclic rings

2.1 Types of stereoisomerism

Before discussing the behaviour of alicyclic compounds it is useful to review several aspects of molecular stereochemistry which will be important later. For a given molecular formula various types of isomerism are possible. Firstly the atoms may be assembled in a variety of different ways (paying attention to the normal rules of valency!). This leads to a number of alternative chemical structures having different atom-atom connectivity patterns; these are known as *structural isomers*. In general however a given structural arrangement may form additional isomers having the same connectivity but differing in the spatial arrangements of the atoms within the molecule; these are known as *stereoisomers* and fall into two classes: *enantiomers* and *diastereoisomers* (sometimes called diastereomers).

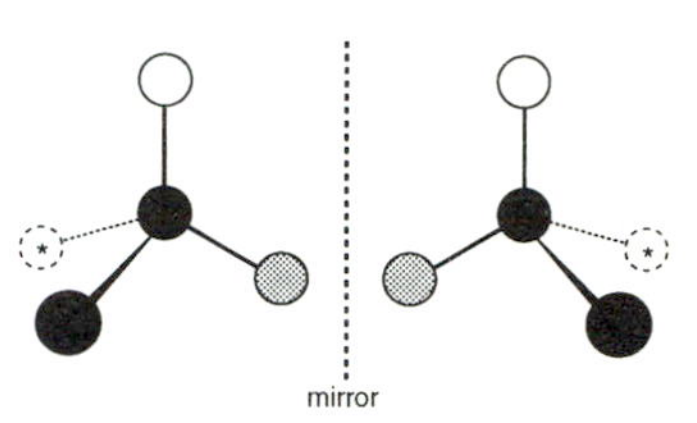

These mirror-image structures are a pair of enantiomers and cannot be superimposed on each other.

Enantiomers occur in pairs and may be most simply defined as structures which are non-superimposible mirror images of each other. All other stereoisomers derived from a particular structural formula are referred to as diastereoisomers. These latter include, for example, *E*- and *Z*-alkenes (often called geometric isomers) or any optically active isomers which are not related as enantiomers.

A consideration of isomerism in 1,2-dichlorocyclopentane illustrates the application of these concepts (it may be useful for the reader to construct models of these). Three different stereoisomers are found as shown below:

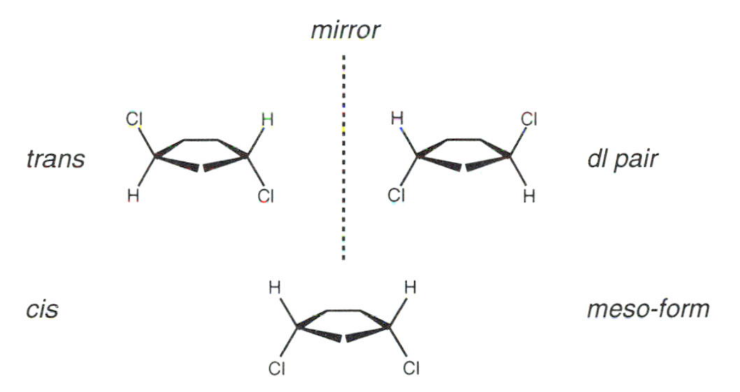

In this case the two chlorine atoms may be located *cis* or *trans* with respect to the ring plane (the *Z,E* system of nomenclature is not used for cyclic compounds). The *cis* isomer has a plane of symmetry, but the *trans* isomer exists as a pair of enantiomers. Such a situation is generally true in 1,n-disubstituted cycloalkanes with two identical substituents. However, in even- membered rings when the substituents are at opposite corners of the structure neither isomer is optically active since in this case each isomer has a plane of symmetry along the line joining the two substituted atoms in addition to other elements of symmetry whether or not the substituents are the same.

In general 1,n-disubstituted cycloalkanes bearing two identical substituents (n being any other position about the ring, such that n > 1) have an achiral *cis* isomer (a meso form) and a chiral *trans* isomer (*i.e.* a pair of enantiomers). These *cis* and *trans* stereoisomers are diastereomers and are often referred as geometric isomers.

In some cases such as 1,4-disubstituted cyclohexanes the *cis* and *trans* isomers are both achiral:

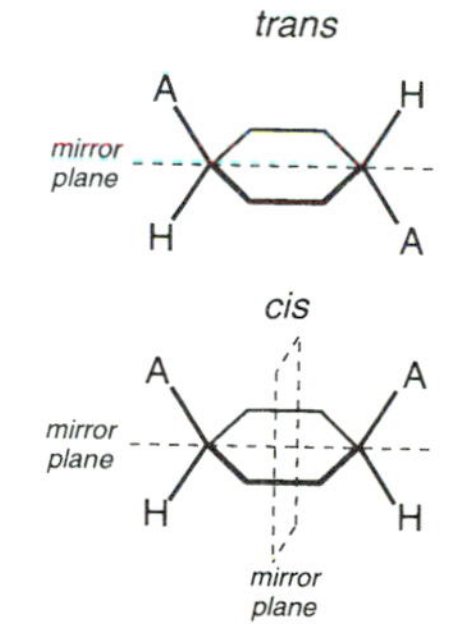

Note that the *trans* isomer also has a centre of symmetry and that, in such flexible ring structures, the stereochemical relationship between different groups is most easily determined by consideration of a planar ring geometry (*if* that can be formed).

2.1.1 Nomenclature When there are more than two substituents present in a ring the nomenclature must be extended to include the relative positions of each group. In such cases it is necessary to define a reference substituent relative to which the locations of the others can be referred. The reference group, which is designated by the prefix *r*, is chosen as that attached to the lowest numbered ring atom bearing a substituent which gives rise to cis-trans isomerism. Other substituents can then be described as *c* (cis) or *t* (trans) with respect to this reference position. Using these rules the compound shown below can be identified as *t*-5-chloro-*t*-2,*c*-3-dimethylcyclohexan-*r*-1-ol.

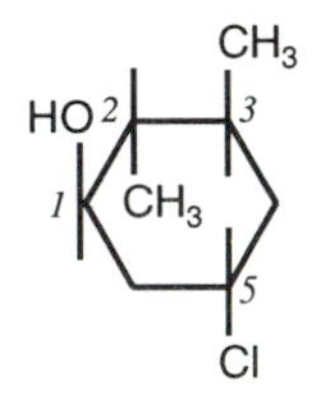

2.2 Configuration versus conformation

It is important to emphasise the difference between configurational and conformational isomerism. Different molecular configurations, *e.g.* enantiomers or diastereoisomers, can normally be interconverted only by the breaking of one or more bonds within the structure. However, flexible molecules can adopt a number of different geometries some of which will be more stable than others, a situation which is very common in alicyclic compounds. Such different molecular geometries which can be readily interconverted by bond rotation are known as conformers or conformations.

[In practice, the difference between configurations and conformations is merely a matter of activation energy. There is a grey area when the activation energy for interconversion is such that the process can occur at a reasonable rate not far above room temperature.]

2.3 Ring conformations

Baeyer related strain to the internal bond angles in planar rings.

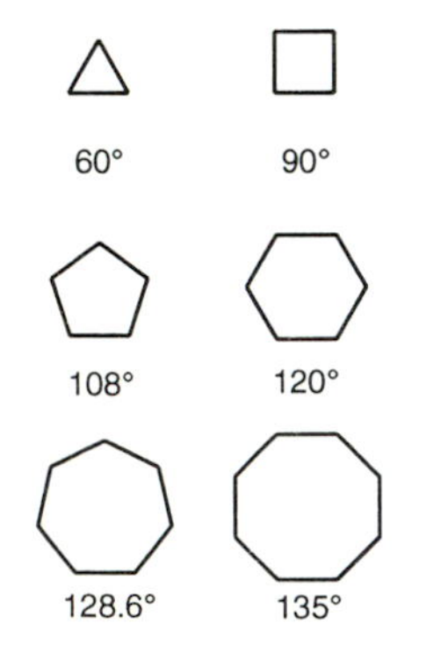

The earliest attempt to explain the behaviour of alicyclic compounds was made by Baeyer whose *ring strain theory* of 1885 assumed that rings were planar and that strain simply reflected deviation from a "normal" tetrahedral carbon bond angle *i.e.* 109° 28' (as seen in methane, diamond etc.). If rings are considered to behave as regular planar polygons it will be seen that cyclopentane with an internal angle of 108° can most readily accommodate sp^3-hybridised carbon atoms; consequently this should be the least strained ring.

Unfortunately this approach breaks down for larger rings because of their non-planarity (even cyclohexane was at that time thought to be planar) but angle strain is none-the-less one of several factors which control ring stability.

2.3.1 Methods for investigating molecular conformation and strain

A number of different techniques have provided valuable insight into the conformational behaviour of saturated carbocyclic rings. Single crystal X-ray structural studies provide direct insight into 3-dimensional geometries in the solid state but the results must be treated with caution in the case of flexible structures since crystal lattice/packing requirements may distort the ring conformation. Consequently such data must be used in conjunction with other approaches. Electron diffraction also provides structural information but this technique uses samples in the gas phase and thus at best only offers a 2-D projection-based view of the structure.

Infra-red spectroscopy gives direct information about bond vibrations and rotations. These depend on vibrational force constants which reflect the hybridisation states of the atoms involved. Increased "s"-character at a carbon atom makes a C-H or C-C bond shorter and stronger (the bond force constants change). This is seen in the C-H infrared absorptions of ethane, ethene and ethyne for which ν_{max} (C-H) changes from 2860 and 2950 cm^{-1} (C_2H_6), to 3010-3060 (C_2H_4) and 3300 cm^{-1} (C_2H_2). The C-H stretching bands for cyclopropane occurs at 3000 and 3062 cm^{-1} suggesting alkene-like character.

Cyclopropanes and unsaturated systems also show absorptions in the uv-visible spectrum. Cyclopropane itself has an absorption band at *ca.* λ_{max} 195 nm, close to that for ethene (at 171 nm) and indicating the relatively weak nature of its C-C σ-bonds (normal C-C σ-bonds show no such ultra-violet absorption).

Of particular importance, however, is the use of NMR (nuclear magnetic resonance) spectroscopy. This can provide direct information about molecular geometry and dynamic behaviour in solution and can also give insight into the hybridisation state and relative geometries of various atoms in a structure.

Chemical shift (δ) and coupling constant (J) data which are obtained from nmr spectroscopy can give useful information about the geometry and conformational flexibility of a molecule in solution, and the hybridisation state of the various atoms in the structure.

Each of these techniques will provide important information in the subsequent discussion.

2.4 Types of strain

Strain arises from a non-ideal molecular geometry and several factors may be considered to make important contributions to it. These include:

- bond angle strain;
- bond length strain;
- torsional strain;
- non-bonded interactions.

When a molecule is in a very unfavourable geometry, its energy can be minimised by distorting bond angles and changing bond lengths (*e.g.* by rehybridisation) but it is not possible to completely compensate for a non-ideal situation. Consequently there remains some residual strain which leads to decreased stability.

In principle it is possible to calculate the degree of strain present in different molecular geometries using an equation of the type:

$$E_{strain} = \sum E(r) + \sum E(\theta) + \sum E(\varphi) + \sum E(d)$$

where:

$E(r)$ refers to bond length strain (*i.e.* bond stretching or compression);

$E(\theta)$ represents a bond angle distortion term;

$E(\phi)$ reflects torsional strain arising from rotation about bonds (ϕ is the dihedral angle between the interacting groups);

$E(d)$ is included to cover non-bonded interactions between groups or atoms (van der Waals repulsion effects).

Each of these factors will be discussed in turn.

The Morse curve.

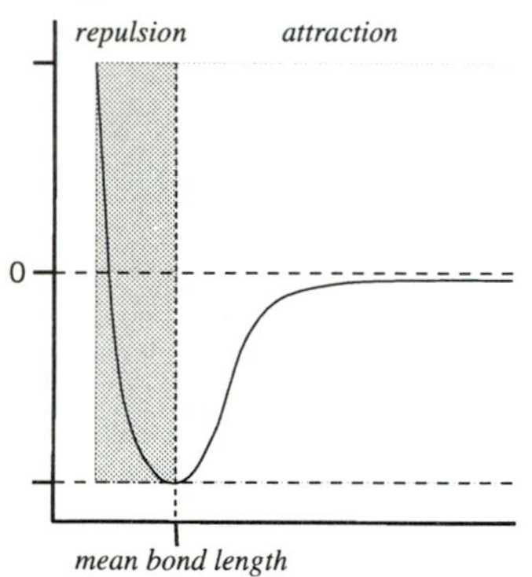

2.4.1 Bond length strain The variation of potential energy of a single bond depends on internuclear distance r in a manner expressed by the Morse curve and approximates a parabola at the bottom of the well. In this region the bond energy therefore varies as $(r - r_0)^2$ where r_0 is the mean bond length in question:

$$E(r) = \frac{1}{2} k_r (r - r_0)^2$$

For a standard C-C single bond with a bond length *ca.* 153 pm (r_0 is usually given the value 152 pm) the constant in this expression k_r has a value *ca.* 26 kJ mol^{-1} pm^{-2}. Therefore deforming a C-C or C-H single bond by as little as 0.3 pm requires *ca.* 1.2 kJ mol^{-1} and a change of bond length is energetically very costly.

2.4.2 Bond angle strain The idealised tetrahedral bond angle is 109° 28' and deviation from this optimum value through bond bending will introduce strain into the structure. The energy change $E(\theta)$ introduced by bond angle deformation is given by the equation:

$$E(\theta) = \frac{1}{2} k_\theta (\theta - \theta_0)^2$$

where k_θ is the appropriate force constant [and typically has a value of *ca.* 0.16 kJ mol^{-1} deg^{-2}]. Bond angle change is energetically much less costly than bond stretching (about 1.3 kJ mol^{-1} for a 4° bond deformation) despite the impression given by molecular models (!) and thus plays a major rôle in reducing molecular strain.

Bond angle change may be thought of as arising from rehybridisation of the pivotal atom *i.e.* through variation of the s- and p-content of the various bonding orbitals. This has the "knock-on" effect of changing bond lengths, strengths and reactivity.

2.4.3 Torsional strain Torsional strain (which is sometimes known as Pitzer strain) arises from repulsive interactions between substituents and bonds on adjacent atoms.

For example, in ethane there is free rotation about the C-C bond leading to a number of different conformations. However, these are not all of the same energy. The least stable arrangement occurs when all the bonds are closest together, *i.e.* when all the bonds eclipse each other when viewed from along the C-C bond, whereas rotation by 60° from this *eclipsed* geometry leads to the *staggered* conformation, the lowest energy structure in which all the bonds are as far away from each other as possible. Further rotation by 60° leads back to the eclipsed conformation and so on reflecting the 3-fold rotation symmetry present along the C–C bond.

The angle of twist, the dihedral angle between the two interacting bonds, ϕ, is also known as the *torsion angle* and the interaction causing the variation in energy, $E(\phi)$, is called *torsional strain*.

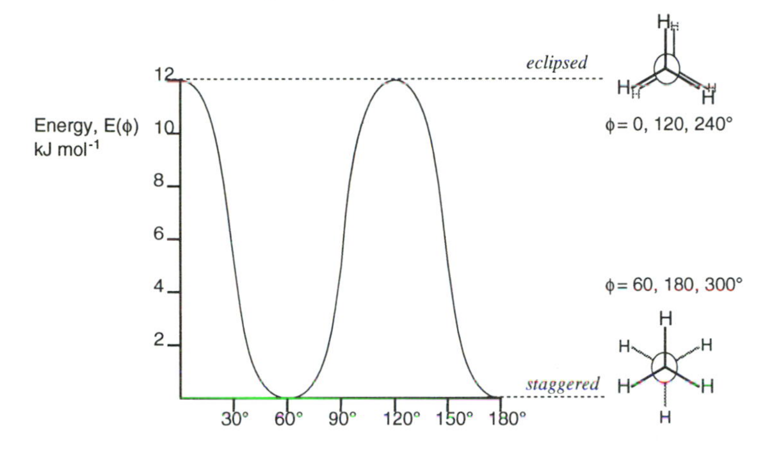

The rotational barrier in ethane

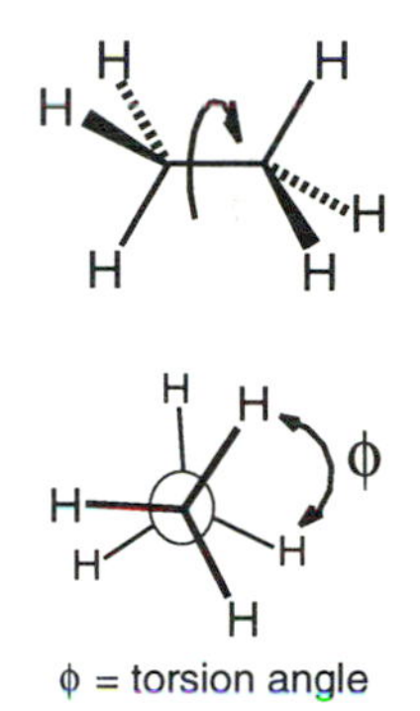

The eclipsed geometry is *ca.* 12 kJ mol^{-1} higher in energy than the staggered situation and since there are three pairs of C-H bonds involved here, each eclipsing C-H interaction can be estimated to raise the energy of the structure by *ca.* 4 kJ mol^{-1}. The energy profile shown above illustrates how the stability of the ethane molecule varies with rotation about the C-C bond. The hydrogen atoms in ethane are too far apart for direct van der Waals' repulsion; the origin of this torsional strain is to be found in electronic interactions within the molecule.

2.4.4 Non-bonded interactions. As the name implies this term relates to the energy cost involved when groups not directly bonded to each other are pushed too close together. The potential energy curve (the Lennard-Jones potential) has a form similar to that of the Morse curve. When non-bonded atoms are well separated, they are weakly attracted to each other by London or dispersive forces through the "mutual polarisation of each others electrons". The potential energy of the system slowly falls to a minimum at an internuclear distance corresponding with the sum of the van der Waals' radii. Further reduction in the internuclear separation leads to a rapid rise in energy

Non-bonded interactions arise from close contact between non-bonded groups within the molecule which are too far apart for their interactions to be accounted for as torsional interactions.

(*i.e.* repulsion) because of interpenetration of the non-bonded electron clouds.

2.4.5. Molecular mechanics calculations Much effort has been put into developing mathematical functions ("force-fields") which accurately model the different strain interactions described above for a wide variety of molecules. These allow the strain energy of any particular conformation to be estimated, and by calculating the different energies of various molecular geometries it is possible to predict the lowest energy conformation. If appropriate force-fields are used such "Molecular Mechanics" calculations can accurately model the behaviour of alicyclic molecules.

2.5 Rotation in butane

When larger groups are present the torsional energy profile, which has just been discussed for ethane, is significantly distorted. In butane, for example, rotation about the central C–C bond introduces two additional factors which must be considered, namely (i) interactions between the terminal methyl groups and (ii) those between a methyl group attached to C_2 and the hydrogens on C_3 (or *vice-versa*).

During a 180° rotation about the central C_2–C_3 bond a number of energy maxima and minima are encountered. Using the dihedral angle ϕ between the two C–Me bonds (see below) to define the molecular geometry it is clear that the lowest energy conformation has the two methyl groups as far away from each other as possible with all bonds staggered (ϕ = 180°). This orientation is known as the anti-periplanar geometry. The highest energy geometry has ϕ = 0°; all bonds are eclipsed and the two methyl substituents are as close as possible – this is the syn-periplanar conformation. Two other key points occur during rotation, at ϕ = 60° and ϕ = 120°, and are known as the syn-clinal, or gauche, and the anticlinal conformations respectively. The gauche conformation is an energy minimum (all bonds staggered) whereas the anticlinal geometry is an energy maximum (all bonds eclipsed) on the rotational energy profile.

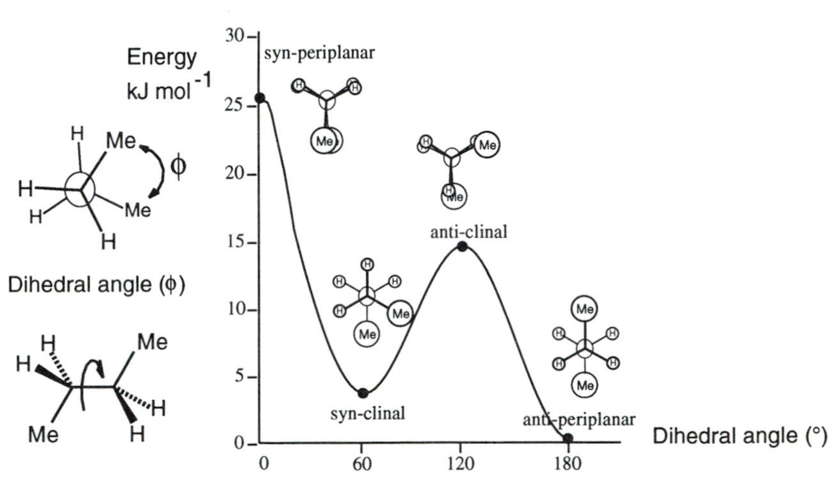

The relative energies of the different conformations are shown above. A C-methyl-C-methyl eclipsing interaction raises the energy by *ca.* 17 kJ mol^{-1} (the syn-periplanar geometry is about 25 kJ mol^{-1} higher in energy than the anti-periplanar form but allowance must also be made for two HC-CH eclipsing interactions, each costing *ca.* 4 kJ mol^{-1}, which are present in the syn-periplanar conformation) whereas a MeC-CH eclipsing interaction costs *ca.* 5.5 kJ mol^{-1} (the anticlinal conformation is *ca.* 15 kJ mol^{-1} higher in energy than the antiperiplanar situation; this value includes one HC-CH and *two* MeC-CH eclipsing interactions). The corresponding MeC-CMe gauche interaction (where $\phi = 60°$) destabilises the structure by *ca.* 3.8 kJ mol^{-1}. [*n.b.* these values are sensitive to the phase (solution or gas) under consideration]

It is important to appreciate that the two eclipsing interactions noted here (MeC-CMe and MeC-CH) retain the electronic component discussed for HC-CH in ethane, but are augmented by a non-bonded steric (van der Waals) repulsion.

Rotation about either C–C bond in propane is much less complex since each eclipsed conformation involves two CH-CH and one HC-CMe interaction. The presence of this latter is reflected in the slightly raised energy barrier (*ca.* 14.2 kJ mol^{-1}) for this rotation process in comparison with that for ethane, the additional 2 kJ mol^{-1} presumably being a measure of CH_3/H van der Waals repulsion in this structure.

2.6 Some other factors influencing conformation

We have so far only discussed the conformational behaviour of hydrocarbons in which electrostatic interactions between groups within the molecule are negligible. However, this situation is no longer true when polar substituents such as chlorine atoms are introduced into the molecule. The presence of dipolar bonds means that the overall molecular dipole moment is conformation dependent. For example the "anti-gauche" energy differences in gaseous 1,2–dichloroethane and 1,2–dibromoethane are *ca.* 4.5 and *ca.* 6.5 kJ mol^{-1} respectively.

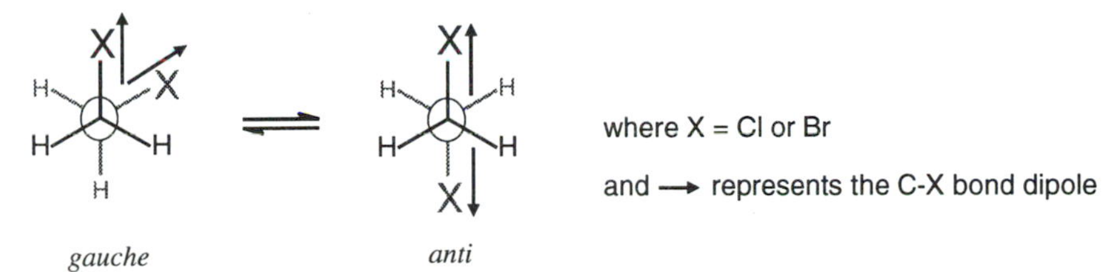

In both cases the energy difference is appreciably larger than that seen for butane despite the fact that the halogen substituents are significantly smaller than methyl groups. This implies strong dipole-dipole repulsion in the gauche conformation. The effect of this repulsion is significantly reduced in polar solvents since the conformation with the higher dipole moment gains more solvation energy under such conditions, thereby increasing the population of the gauche conformer.

Another situation where the interaction between polar groups can play an important rôle in determining conformational preferences and thereby influencing reactivity is in structures of the type CH_3O-CH_2X where X = halogen, OR, OCOR, SR or NR_2). In general in such cases it is found that a gauche conformation is preferred. In this situation the substituent X will be located antiperiplanar to an oxygen lone pair. In general there is a stereoelectronic preference for conformations in which the best donor lone pair or bond is placed antiperiplanar with respect to the best acceptor bond. This is known as the *anomeric effect* and plays an important rôle in sugar chemistry.

Originally this phenomenon was explained in terms of dipole-dipole repulsion between the C-X bond and the C-O-C subunit but the energetics of this interaction would not be enough to account for the magnitude of the anomeric effect. More recently, anomalies which have been noted in the bond lengths of molecules of the type MeO-CH_2-X have led to the suggestion that the anomeric effect arises from an interaction between a non-bonded (lone-pair) orbital on oxygen and the C-X σ^* antibonding orbital. This would strengthen the MeO-CH_2 bond and weaken the CH_2-X bond.

The *Anomeric Effect* can be considered as resulting from an n-σ^* interaction between a C-X bond and a lone pair on an adjacent heteroatom.

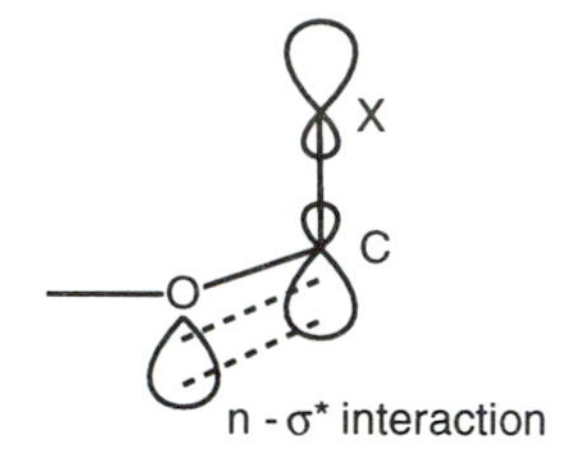

Gauche conformations can also be stabilised by other interactions such as intramolecular hydrogen bonding. The infrared spectra of the meso- and (d,l) isomers of butane-2,3-diol each show two distinct O-H bands, one due to free -OH, the other intramolecularly hydrogen-bonded OH. The band separation ($\Delta\nu_{max}$) is 42 cm^{-1} in the meso and 49 cm^{-1} for the (d,l) isomer. The larger separation observed for the (d,l) structures indicates more favourable intramolecular hydrogen bonding in this compound. For the (d,l) isomer hydrogen bonding increases the torsion angle between the C-Me bonds thereby lowering the energy of the structure whereas for the meso case this angle is reduced increasing torsional interaction between the methyl groups and raising the energy of the hydrogen bonded situation.

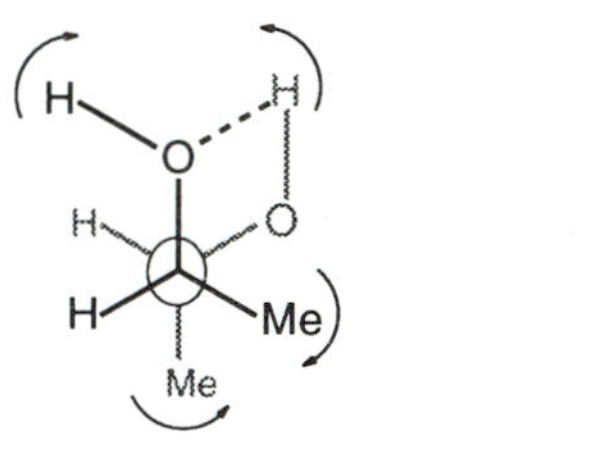

meso-isomer

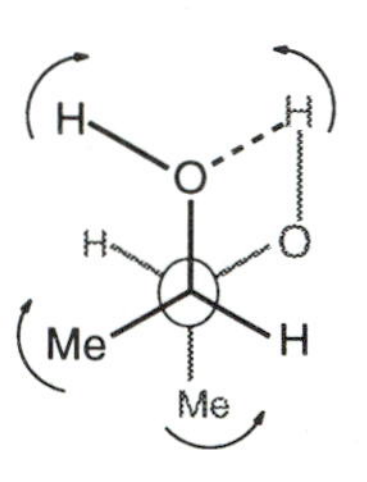

d,l-isomer

2.7 Measurement of strain and classification of ring sizes

2.7.1 Quantification of Strain Examination of the Heats of Combustion (ΔH_{comb}) for a series of saturated carbocyclic rings having between 3 and 14 carbon atoms reveals that there is a marked dependence on ring size. These values (as ΔH per carbon atom) are listed in the Table below but are more readily interpreted using the bar chart accompanying it. The lowest enthalpy change, corresponding to the most stable ring, is found for cyclohexane (n = 6) and this value can be used as a reference point for analysing the remainder. The difference (Δ_n/n) between these [where $\Delta_n = \Delta H_n - \Delta H_6$] is plotted in the Figure. Any ring having a Heat of Combustion (ΔH_{comb}) per carbon atom greater than that for cyclohexane is less stable than cyclohexane and may therefore be regarded as having ring strain. This is a concept which will be returned to shortly.

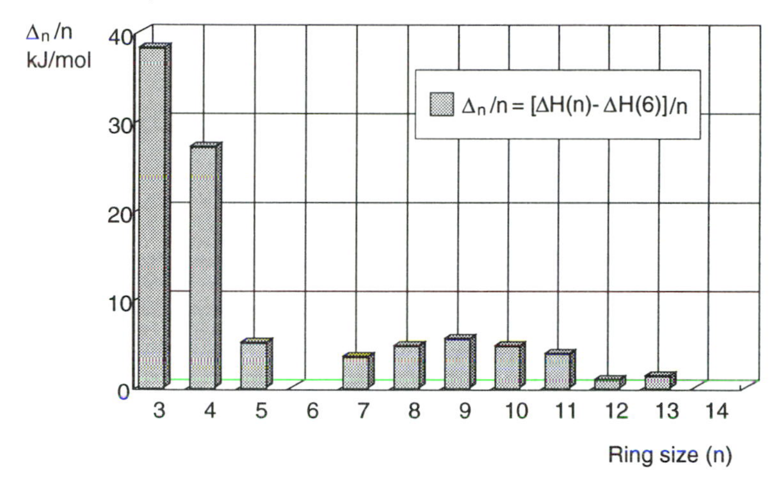

Estimation of strain in alicyclic rings (relative to cyclohexane)

This bar chart shows the strain per carbon atom Δ_n/n present in alicyclic rings relative to cyclohexane.

Relative Heats of Combustion for alicyclic rings.		
Ring Size	ΔH/n	Total Strain
3	38.6	115.8
4	27.4	109.6
5	5.4	27.0
6	0.0	0.0
7	3.8	26.6
8	5.0	40.0
9	5.8	52.2
10	5.0	50.0
11	4.2	46.2
12	1.2	14.4
13	1.6	20.8
14	0.0	0.0
	kJ/mol	kJ/mol

From this plot it is clear that there is a marked difference in stability between different ring sizes and we shall see that this has important consequences for many aspects of the behaviour of these compounds. Indeed this plot or its inverse frequently indicates the behavioural trends observed for alicyclic derivatives and it is conveniently summarised by the block diagram shown on the right.

Classification of alicyclic rings based on the degree of strain:

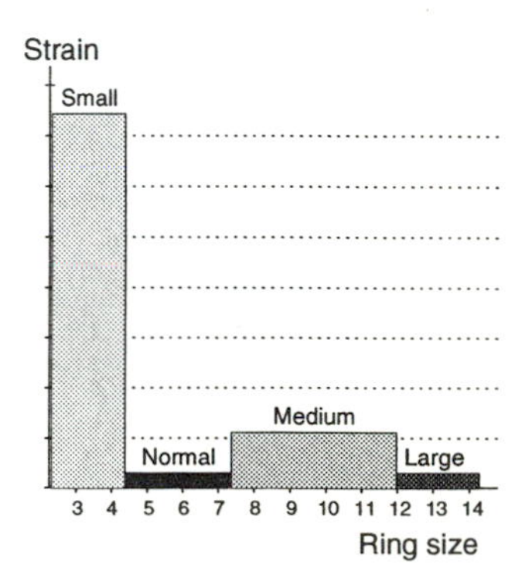

2.8 Classification of alicyclic rings

The block diagram suggests that alicyclic rings can be conveniently divided into four main groups: small (n = 3 and 4); normal (n = 5-7); medium (n = 8-11); and large ($n \geq 12$). This classification is valuable when discussing their preparation, properties and reactivity.

2.9 The consequences of strain in small rings

2.9.1 Cyclopropane The three carbon atoms of cyclopropane are necessarily coplanar and therefore the molecule must have a nominal bond angle of 60°. Consequently such a structure should be highly strained. However, the strain difference between cyclopropane and cyclobutane is much smaller than would be expected on the basis of bond angle changes and other effects must help stabilise the three-membered ring. It is useful to compare the physical properties of cyclopropane with corresponding data for ethane, ethene and ethyne:

	cyclopropane	ethane	ethene	ethyne
H-C-H angle,°	118	109.5	120	-
r(C-H), pm	108	110.3	108.6	105.7
r(C-C), pm	154	154	134	120
ν(C-H), cm^{-1}	3000, 3062	2960, 2860	3010-3060	3300
H-C-C angle,°	116.4	109.5	120	180
		→ increasing "s"-character in C-H bonds →		

These data suggest that cyclopropane behaves more like an alkene than a saturated hydrocarbon. In particular $\nu_{C\text{-}H}$, $r_{C\text{-}H}$ and the H-C-H bond angle in cyclopropane and ethene are very similar; only the C-C bond length $r_{C\text{-}C}$ in cyclopropane is like that of ethane. A change from sp^3 to sp^2 hybridisation leads to increased "s"-character at carbon in the C-H bonds and consequently increased "p"-character for the C-C bonds in cyclopropane. Since p-orbitals are 90° apart, such a situation more nearly fulfils the requirements for a three-membered ring geometry. Increased "p"-character in the C-C bonds does not lead to a reduction in bond length but does result in reduced bond strength as is evident from the chemical behaviour of cyclopropane.

It is thought that the electron density of the cyclopropane C-C bonds does not lie directly along the C-C bond axis but rather is splayed out in a manner which effectively widens the C-C-C bond angle. Such σ–bonds are known as "banana" bonds.

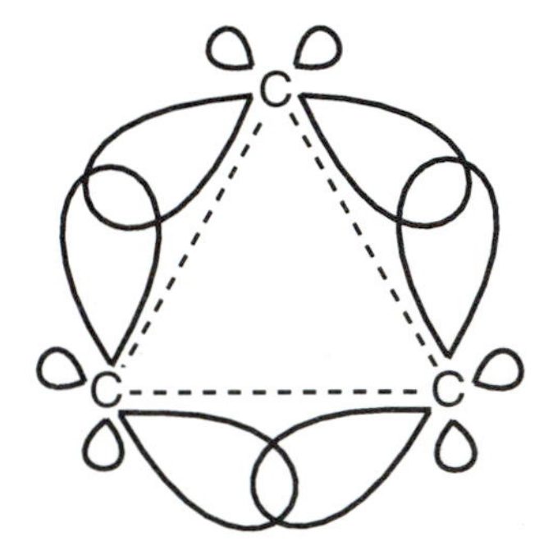

Most of the strain in cyclopropane arises from inefficient overlap of orbitals weakening the carbon-carbon bonds in the structure. From a chemical perspective, there is a continuous gradation in behaviour from a normal alkene like ethene via cyclopropane to cyclobutane (which behaves like an alkane). Despite their reactivity, cyclopropane derivatives are common and even occur in natural products *e.g.* the pyrethrins, compounds with insecticidal properties which are found in certain members of the chrysanthemum family.

A pyrethrin:

2.9.2 Cyclobutane This molecule is thought to have an essentially classical structure. A planar ring conformation suffers from minimum angle but maximum torsional strain because of eight pairs of C-H eclipsing interactions. These latter can be reduced by ring puckering (at the expense of some increase in angle strain) and electron diffraction and X-ray data on cyclobutane and its derivatives do indeed show a wing-shaped molecule having an angle of fold θ of ca. 30° and a barrier for ring inversion in the parent molecule of 6 kJ mol^{-1}.

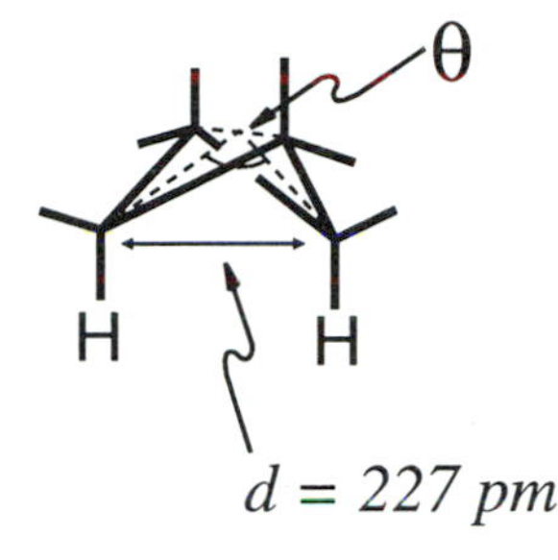

Cyclobutane has $r_{C\text{-}C}$ = 156 pm, $r_{C\text{-}H}$ = 109 pm, and a C-C-C bond angle of 104°, values which support the view of a relatively classical alkane-like structure, but it is none-the-less rather unstable - why? Cyclobutane remains a structural compromise, puckering relieves torsional strain but increases bond angle strain and brings transannular carbon atoms close together resulting in significant non-bonded repulsion.

2.9.3 Use of NMR coupling constants to probe molecular structure The $^{13}C\text{-}^{1}H$ nmr coupling constant is sensitive to the electronic state of the carbon atom to which the proton is attached. For example the change in the magnitude of this coupling in methane (J = 125 Hz), ethene (J = 157 Hz) and ethyne (J = 248 Hz) reflects the change from sp^3 to sp hybridisation. The variation of J_{CH} in the cycloalkanes (shown right) clearly reflects increased "s"-character in the small rings, particularly cyclopropane, and has been used to deduce that the C-C bond of cyclopropane has 17% "s"-character (equivalent to sp^5 hybridisation) whereas the C-H bond has 33% "s"-character (i.e. involves an sp^2 hybrid).

$J\{^{13}C\text{-}^{1}H\}$ values for carbocyclic rings:

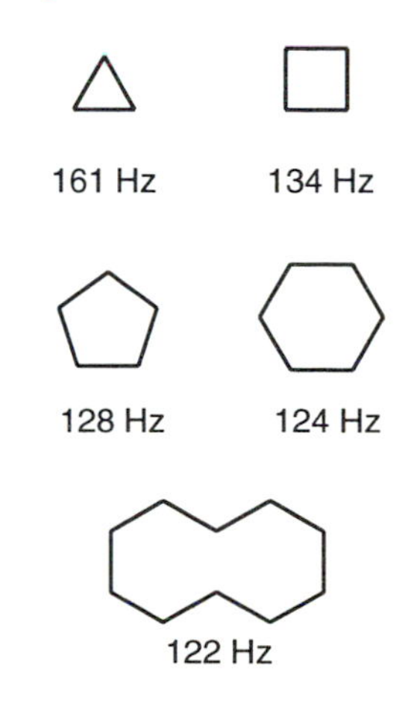

2.10 Conformational behaviour of normal rings

2.10.1 Cyclopentane Cyclopentane is a floppy ring of ill-defined structure. Heat of combustion data suggest that it is somewhat strained despite the near-perfect bond angle present in a planar five-membered ring. The problem is that in the planar geometry all the CH_2 groups are eclipsed introducing *ca.* 40 kJ mol^{-1} of torsional strain. However, the experimentally measured destablisation is rather less than this (27 kJ mol^{-1}) reflecting strain relief brought about by twisting of the ring. X–ray crystallography indicates a C-C-C bond angle of 105°, *i.e.* less then tetrahedral.

There are two key geometries for cyclopentane, the envelope (which has C_s symmetry) and the twist or half-chair form (C_2 symmetric), the envelope being the more stable by *ca.* 2 kJ mol^{-1}. Neither of these is a static conformation; there is rapid fluctuation between them. In this process every atom within the ring spends some time in each of the various different possible locations. There is effectively a continuous up-and-down flexing of the ring which equivalences all the atoms of each type - a process known as "pseudorotation".

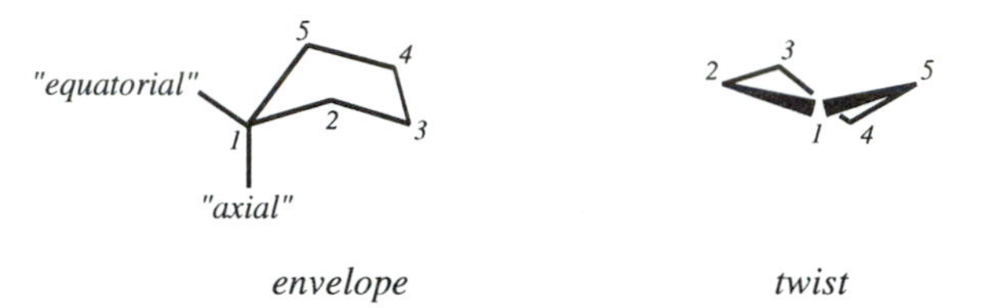

The presence of substituents distorts the energy profile, favouring a particular geometry, indeed the ring may adopt a conformation somewhere between the two extremes. For methylcyclopentane the most stable conformation (by 3.8 kJ mol^{-1}) has the methyl group in an equatorial position (C_1 above) on the flap of the envelope whereas cyclopentanone favours the twist form with the carbonyl group at C_1 (above).

In general, for 1,2-disubstituted cyclopentanes the *trans* isomer is more stable than the *cis*, the major factor destabilising the latter being direct interaction between the substituents themselves, whereas for 1,3–disubstituted derivatives the *cis* isomer usually appears to be the more stable.

In the solid state the preferred conformation is very sensitive to intermolecular interactions. For example, a study of a racemic mixture of *trans*-cyclopentane-1,2-dicarboxylic acid reveals a twist geometry whereas that of the optically active acid shows an envelope conformation. Clearly packing forces can readily perturb the potential energy surface in such conformationally flexible structures!

2.10.2 Cyclohexane The conformational analysis of the cyclohexane molecule is well understood and there have been very detailed studies of many of its derivatives. Examination of a simple molecular model of cyclohexane reveals that, in order to satisfactorily accommodate a tetrahedral bond angle for each ring carbon atom, the structure must pucker - the ring is ***not*** planar. There are three conformational energy minima: the chair, the boat, and the twist-boat (or skew) conformation:

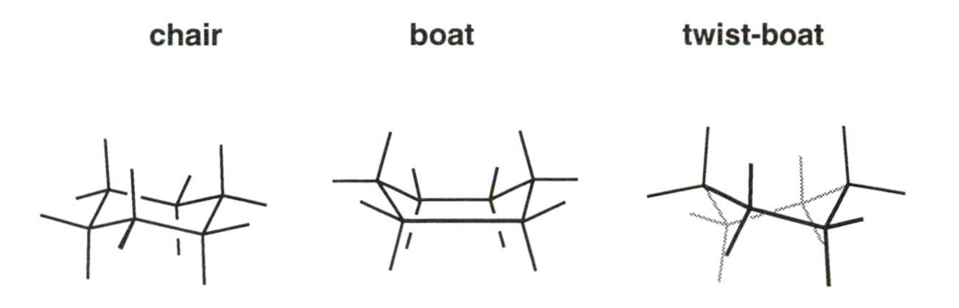

The chair conformation, which has the lowest energy, is rigid and has an almost ideal geometry. For example, it is the key structural building block of the diamond lattice. It is free of angle strain and there are no eclipsing interactions, the structure being composed entirely of gauche butane-type units (each carbon-carbon bond in the cyclohexane ring can be thought of as a butane fragment since each ring carbon atom carries two hydrogens and one saturated carbon atom). The terminal carbon atoms of each "butane-like" ring segment are involved in a gauche-butane interaction as shown below. Two of these are seen in an end view of the cyclohexane chair (there are four more to be found when the ring is rotated - one about each carbon-carbon σ-bond of the ring):

The diamond lattice is built up from chair cyclohexane building blocks.

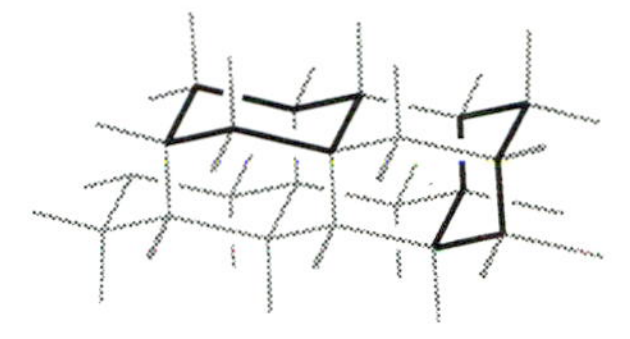

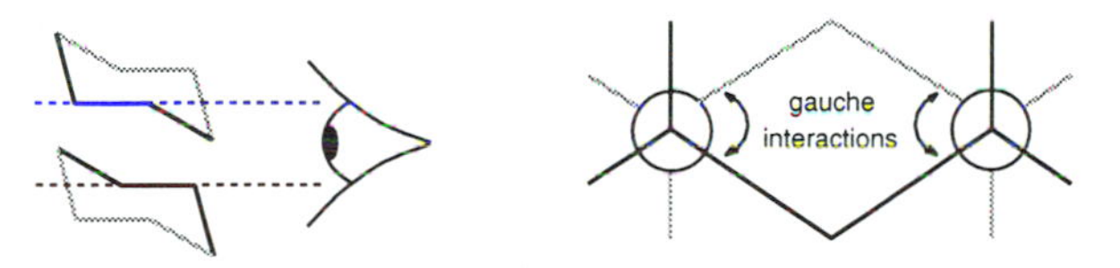

In the cyclohexane chair there are two distinct locations for substituents; these are known as the axial and equatorial positions. That two such sites exist is readily demonstrated by a variable temperature ^{1}H nmr spectroscopic study of cyclohexane-d_{11} since at low temperatures (<-90°C) two resonances of equal intensity are observed reflecting the two different hydrogen environments in the ring. However, when the sample is warmed to room temperature only one absorption is seen; all the hydrogens have become equivalent.

VT NMR studies of cyclohexane-d_{11} show the ring inversion.

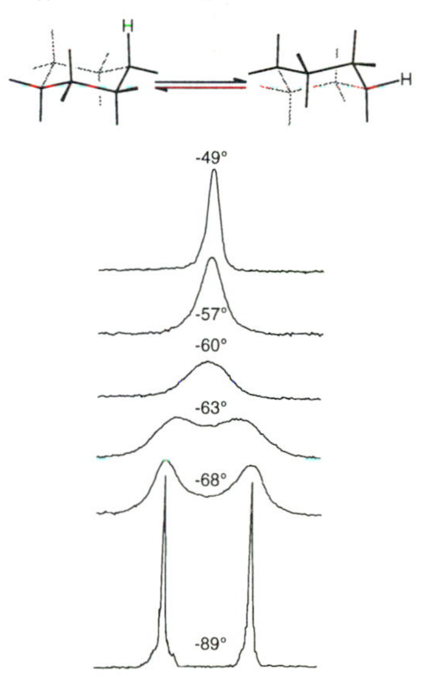

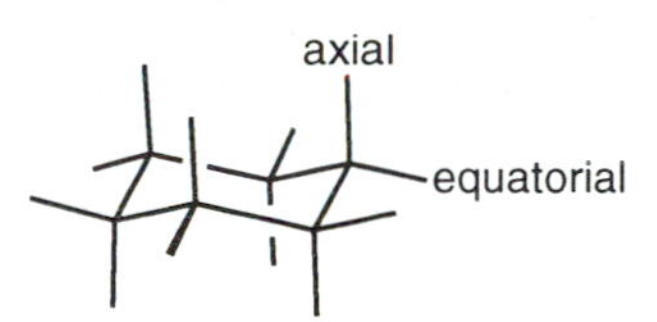

Despite its relatively rigid structure, cyclohexane can readily undergo a ring-flip into a relatively inverted chair geometry in which each of the axial and equatorial protons has changed position. This process occurs fast enough on the nmr time-scale at room temperature to make all of the protons appear equivalent but, at lower temperatures, it is slowed sufficiently that the ring appears to be frozen on the nmr time-scale. Ring inversion in cyclohexane is fast around room temperature (*ca.* 10^4–10^5 times per second at 300K) and interconverts the axial and equatorial positions.

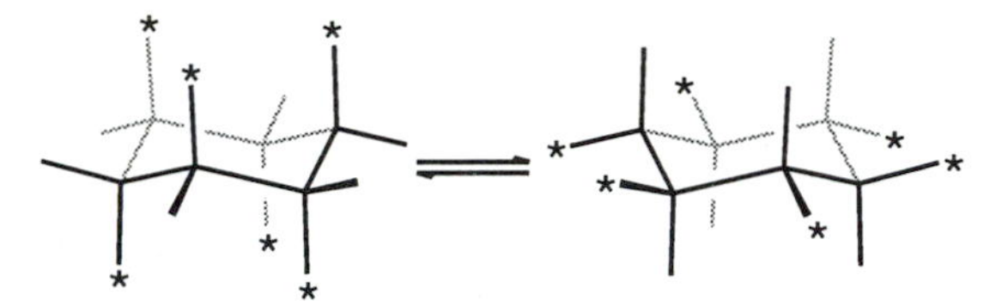

The activation barrier ($\Delta H^\ddagger$) for ring inversion in cyclohexane has been measured using variable temperature nmr data. Probably the "best" values which have been obtained (using d_{11}-cyclohexane) are $\Delta G^\ddagger = 42.9$ kJ mol^{-1} $\{\Delta H^\ddagger = 44.8$ kJ mol^{-1}; $\Delta S^\ddagger = 9.2$ J mol^{-1} K^{-1} at -50 to -60°C$\}$.

Electron-diffraction studies of cyclohexane suggest that, in the gas phase, the chair conformation is slightly flattened from an ideal geometry:

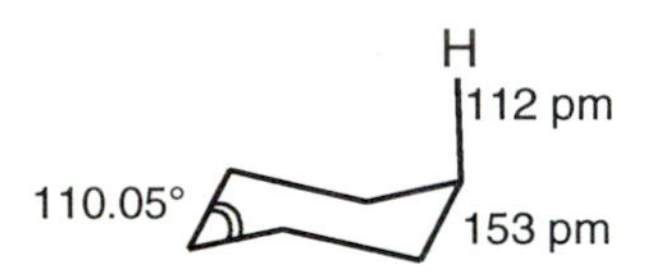

Molecular mechanics calculations have been used to probe the pathway by which this chair-chair inversion occurs. These show that the two boat conformations are of higher energy than the chair (the twist boat being the slightly more stable) and are intermediates in the conversion of one chair geometry into the other.

The transition state through which chair is converted to boat is a twist chair (or half chair) geometry, *ca.* 42 kJ mol^{-1} higher in energy than the chair. In this structure four of the skeletal atoms are coplanar (see below).

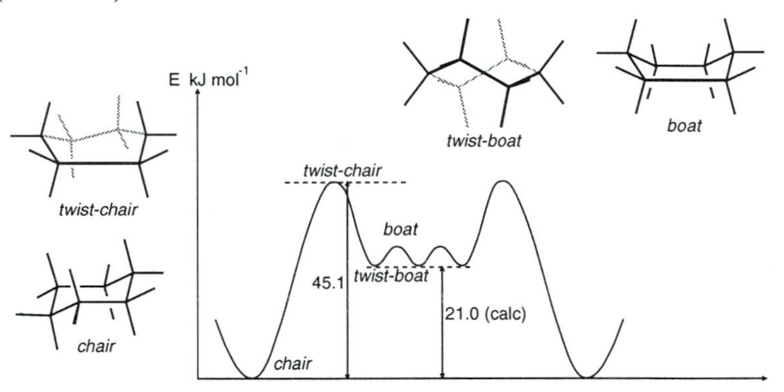

The twist-boat conformation is about 21 kJ mol^{-1} higher in energy than the chair and is 4-6 kJ mol^{-1} more stable than the true boat. The twist-boat conformation is flexible. It lies half way between two relatively inverted boat geometries. Indeed twist-boat conformations can readily interconvert through true boat transtion states.

Direct experimental evidence has been obtained for the existance of the twist-boat. Cyclohexane vapour was passed through a hot tube at 800K and then rapidly condensed onto a CsI plate held at 20K. An infra-red spectrum of the condensed material was then recorded at 70K. This revealed new bands at 770, 869, 1153 and 1469 cm^{-1} which were assigned to the twist-boat which appeared to comprise about 25% of the mixture. The rate of disappearance of the twist boat on slight warming of the mixture gave an activation energy $\Delta G^{\ddagger} = 22.2$ kJ mol^{-1} for reversion to the chair form.

This elegant experiment relies on the fact that whilst the chair form is the enthalpically more stable structure, the twist-boat geometry is favoured entropically (by *ca.* 20 J. mol^{-1} K^{-1}). Consequently, whilst at room temperature (300K) K_{eq} (chair-to-twist) has a value of 4000, at 800K it is *ca.* 3!

Only *ca.* 0.4% of the twist-boat cyclohexane conformer is present at room temperature.

Why should the boat geometries be less stable than the chair conformation? We have seen that the chair conformation of cyclohexane contains six gauche interactions. In the true boat two of these are replace by eclipsed, syn-periplanar interactions with a consequent increase in the energy of the structure.

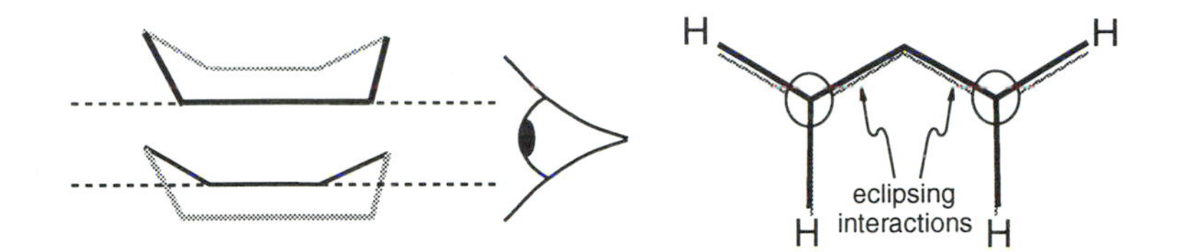

The boat conformation of cyclohexane is of higher energy than the chair because two of the gauche interactions in the latter are replaced by eclipsed (syn-periplanar) interactions.

By analogy with butane the energy difference between a gauche and a syn-periplanar interaction is *ca.* 22 kJ mol^{-1}. This suggests that the two eclipsing interactions in the cyclohexane boat should raise its energy by *ca.* 44 kJ mol^{-1} relative to the chair conformation (a value somewhat higher than those predicted by molecular mechanics calculations and suggested by nmr experiments, the discrepancy being a reflection of the limitations of the butane model).

Non-bonded repulsion across the ring between the "flagpole" hydrogen atoms further reduces the stability of the boat conformation.

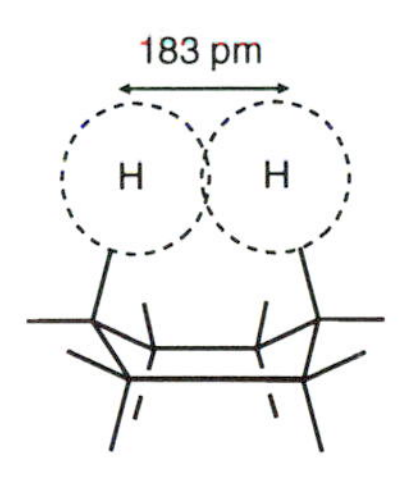

Another factor which destabilises the boat conformation is the non-bonded repulsion introduced by the close proximity of the so-called "flagpole" hydrogens; they are *ca.* 180 pm apart, *i.e.* significantly within the van der Waals' contact distance of 240 pm. In the twist-boat both the eclipsing and non-bonded interactions are relieved as can be seen in the Newman projections shown below:

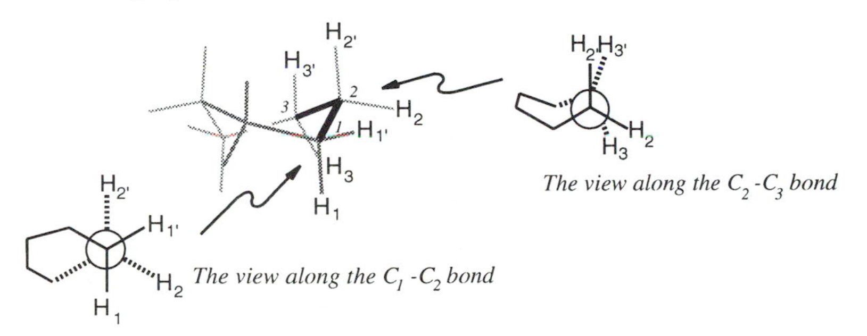

The view along the C_2-C_3 bond

The view along the C_1-C_2 bond

2.10.3 Cycloheptane Rings of more than six carbon atoms tend to be inherently more flexible and can often adopt a number of different conformations of similar energies making conformational analysis more complex.

For example the strain in cycloheptane (25 kJ mol^{-1}) is similar in magnitude to that observed for cyclopentane but in this case there are four main conformations which lie within *ca.* 11 kJ mol^{-1} of each other. These fall into two families: chair and twist-chair; and boat and twist-boat, all of which are flexible. The chair is destabilised by eclipsing interactions at the "flat" end which are relieved by pseudorotation into the twist-chair - the most stable conformation (by 9 kJ mol^{-1}) of cycloheptane. The twist-boat is a little higher in energy (10.4 kJ mol^{-1} above the twist-chair) but is more stable than the simple boat (by 2 kJ mol^{-1}). There is a relatively high potential barrier (36 kJ mol^{-1}) separating the chair/twist-chair and boat/twist-boat conformational families.

There are four main cycloheptane conformations.

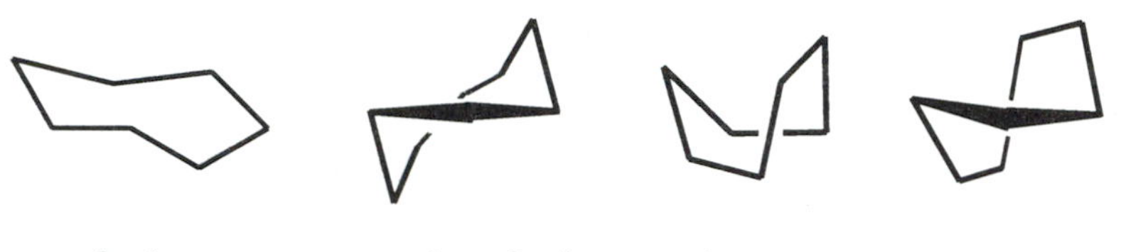

2.11 Medium rings

All the medium rings (C_8-C_{11}) are significantly less stable than cyclohexane. The key sources of strain in these molecules arise from a combination of transannular hydrogen-hydrogen clashing (non-bonded repulsion), bond-angle widening and some torsional interactions.

The total strain in medium rings has sometimes been referred to as "non-classical' strain to distinguish it from simple angle strain (regarded as "classical strain") though this latter is often a substantial contributor to the overall strain in these systems. Whilst transannular van der Waals' repulsion between hydrogen atoms is only a relatively minor contributor to this, it does have other important consequences leading for example to chemical reactions across the ring (see later).

Each of the medium rings have a number of conformations which are relatively close in energy and which rapidly interconvert at normal temperatures. The activation energies for such conformational interconversion are:

Ring size	E_a(kJ/mol)
8	34
9	25
10	25
12	30
14	29

Much insight has been gained about the conformational behaviour of medium ring structures from X-ray structural studies and there is evidence that solid-state conformations are preserved in solution or the liquid state. The lowest energy conformations of even-membered rings appear to be related to the diamond lattice. This is readily seen for the chair-boat-chair, chair-chair-chair and boat-chair-boat conformations of cyclodecane:

Side and top views of cyclodecane in the solid-state.

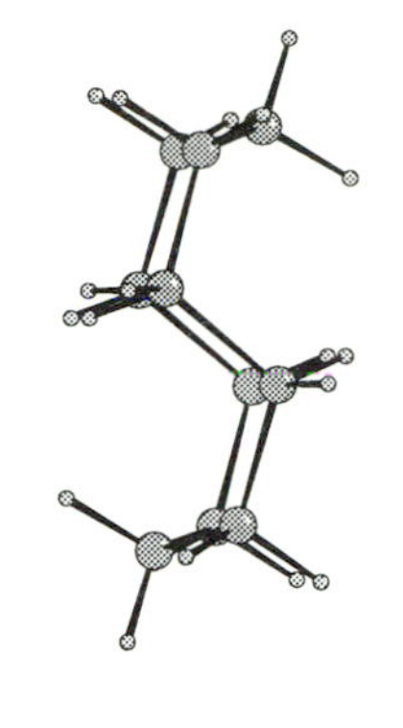

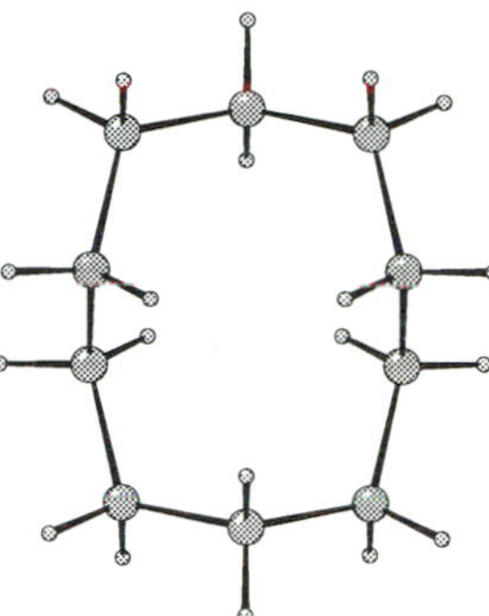

Boat-chair-boat

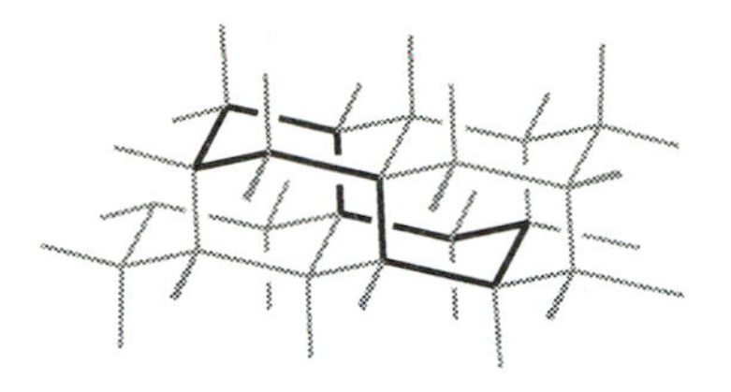

chair-chair-chair

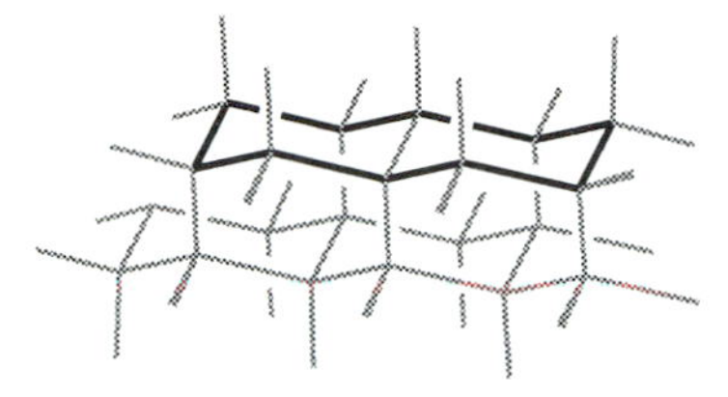

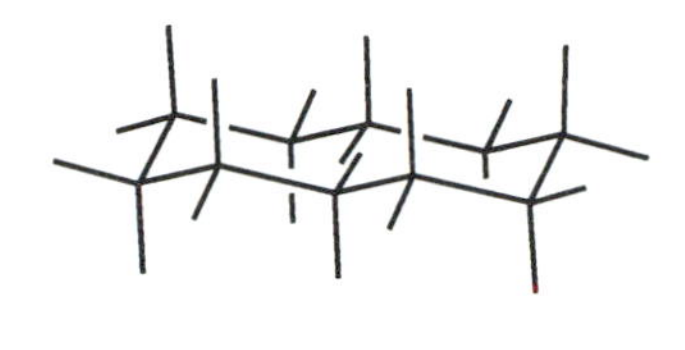

Chair-boat-chair

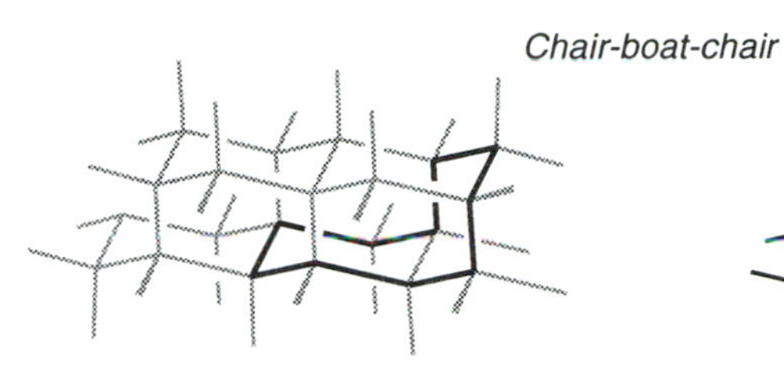

In practice, X-ray diffraction studies of cyclodecane reveal a boat-chair-boat geometry with some bond-angle distortion (angles are significantly wider than tetrahedral) and in which the internal hydrogen atoms are only *ca.* 1.8Å apart.

An alternative view of cyclodecane emphasises this internal interference showing that there are six internal hydrogen atoms in close proximity:

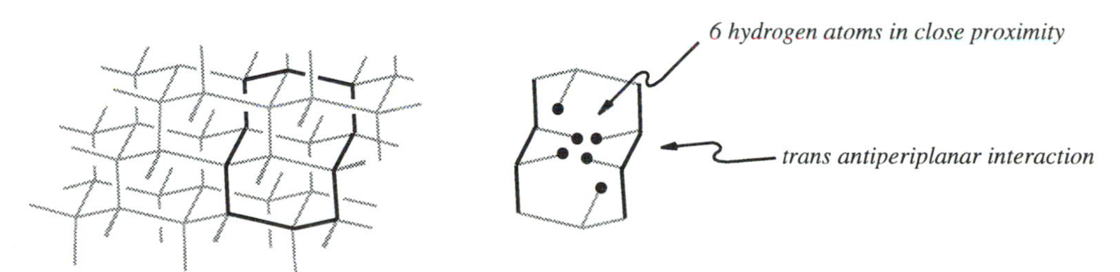

This conformation incorporates two anti-periplanar "butane" interactions (highlighted) but the eight remaining partial butane structures are all skew; these latter add an additional 27 kJ mol^{-1} of torsional strain to the overall energy of the structure.

The "diamond-lattice" approach predicts a boat-boat (or saddle) conformation for cyclo-octane. However, this is not the lowest energy geometry (though it is one of several important contributors) since it is destabilised by hydrogen-hydrogen close contacts:

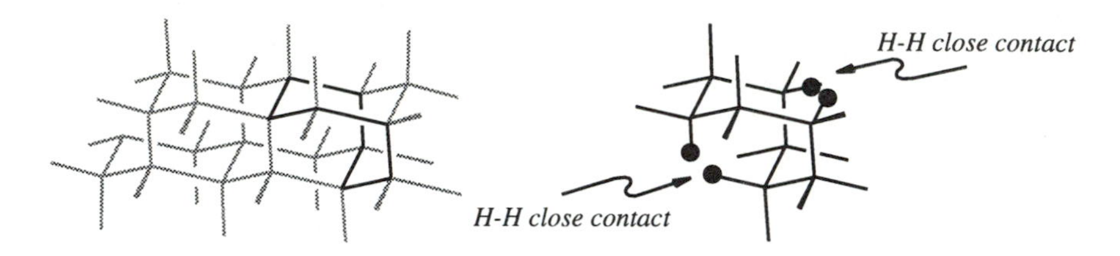

In fact the most stable conformer, and that seen in X-ray structural studies, is the boat-chair though the boat-boat (above) and the chair-chair (or crown) also appear to be important. However, there are eleven different energy minima on the cyclo-octane conformational potential energy surface having an activation energy for interconversion in the range 20-35 kJ mol^{-1}.

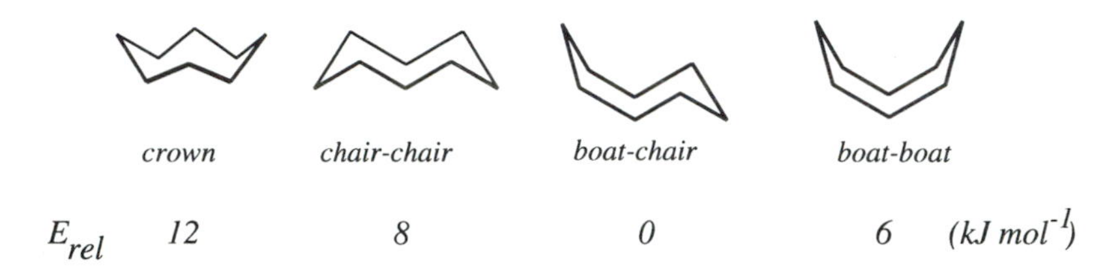

E_{rel}	12	8	0	6	(kJ mol^{-1})

Unfortunately each of the minimum energy conformations has significant transannular repulsions which destabilise the structure. These interactions can be relieved by twisting but this leads to increased torsional interactions!

The diamond-lattice model cannot be applied to the odd-numbered rings (C_9 and C_{11}) which are somewhat less stable than their even-membered neighbours principally because angle strain becomes even more important in these cases.

Top and side views of the "rectangular" conformation found in crystals of cyclotetradecane:

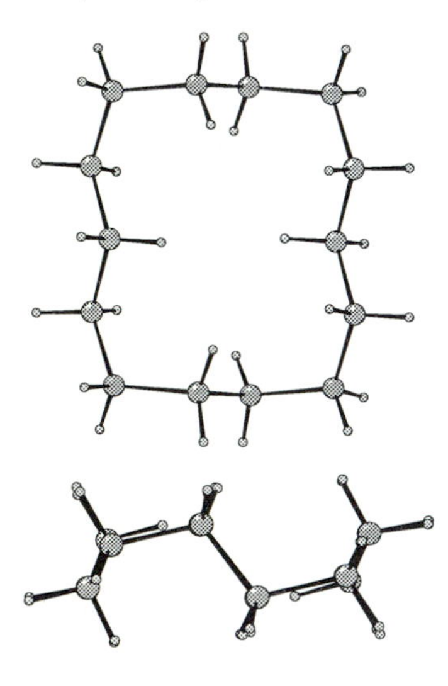

2.12. Large rings

Cyclododecane is slightly strained; it does not fit comfortably into the diamond lattice and X-ray diffraction data shows that it adopts a "square" conformation. In contrast cyclotetradecane and cyclohexadecane prefer conformations based on a diamond-lattice type geometry, "rectangular" for the C_{14} structure and "square" for the C_{16} ring.

Very little information is available about the behaviour of 13- and 15-membered rings.

2.13 Substituted cyclohexanes - the effect of substituents on ring conformation

2.13.1 Mono-substituted rings What happens when a substituent such as a methyl group is placed on a cyclohexane ring? The substituent can, of course, be located either axial or equatorial as in methylcyclohexane shown below, but the two conformations are not of equal stability. The equatorial form is normally the more stable [exceptions occur when X = F or HgBr (which show little conformational preference) and X = HgCl (which slightly favours an axial conformation)].

One axial methyl group

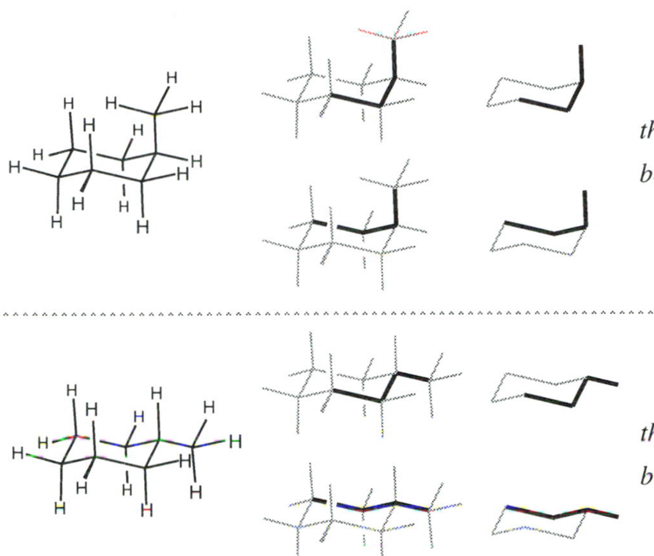

there are two gauche interactions between the substituent and the ring.

there are two antiperiplanar interactions between the substituent and the ring.

One equatorial methyl group

Monosubstituted cyclohexanes undergo rapid conformational inversion at room temperature but the equatorial form is in general the more stable.

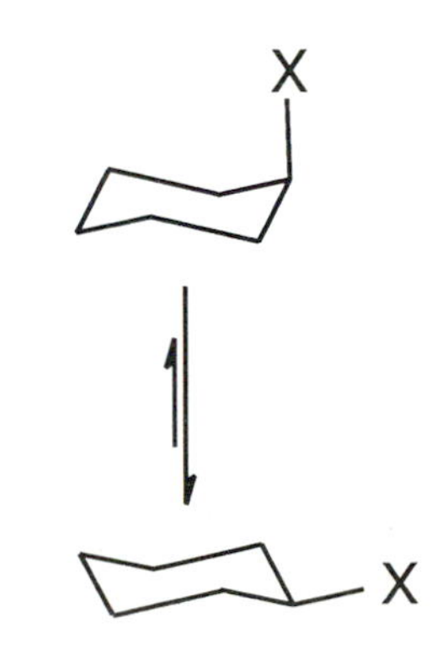

Each of the two gauche interactions in *axial*-methylcyclohexane raises the energy of this conformation by 3.3 kJ mol^{-1} above that of the equatorial conformer. Consequently at room temperature the equilibrium mixture of the two should contain about 95% of the equatorial conformation. However, it has never been possible to separate these two isomers of methyl cyclohexane at room temperature; they are rapidly interconverting. Such interconversion can only proceed through the boat or twist-boat conformation and requires an activation energy of *ca.* 42 kJ mol^{-1}.

Rapid interconversion between the axial and equatorial forms is a general phenomenon for mono-substituted cyclohexanes and to date no conformational isomers of such compounds have ever been separated. Nmr spectroscopic studies (particularly ^{13}C) at low temperatures provide a very convenient method for investigating quantitatively such equilibria (as has already been seen for d_{11}-cyclohexane) but infrared spectroscopy has also been used. Working at very low temperatures pure equatorial conformers of chloro- and trideuteriomethoxy-cyclohexane have been isolated as solids and in solution; indeed the equatorial form of chlorocyclohexane has a half life of 22 years at –160°C but only 0.01 sec at –70°C.

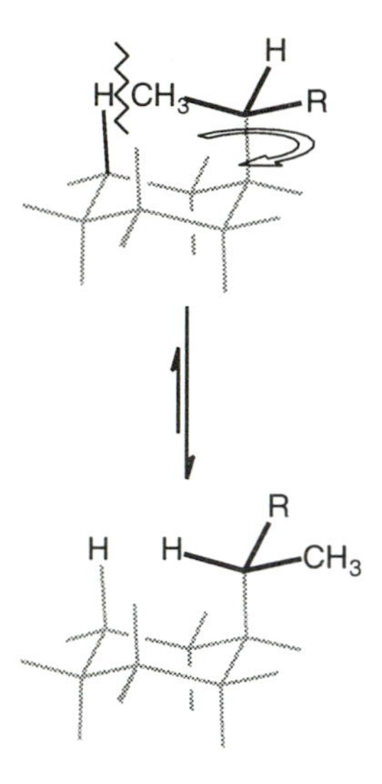

Increasing the bulk of the substituent will increase the energy difference between the axial and equatorial conformers. As the alkyl group is made more bulky the axial-equatorial energy difference increases slightly [$\Delta G° = \sim 7.3$ kJ mol^{-1}(X = Et); ~9.0 kJ mol^{-1} (X= *i*-Pr)]. Rotation of the substituent in the axial geometry relieves steric interactions in the case of an ethyl or an isopropyl group but this is not possible for a *tert*-butyl substituent and in this case the energy difference dramatically increases to $\Delta G° > 20$ kJ mol^{-1} [the axial conformer of *tert*-butyl cyclohexane is sufficiently strained that it has an energy similar to that of the boat conformers and the conformational equilibria involved become difficult to analyse]. Such a value is sufficiently large that only 1 in 10^4 molecules is in the axial conformation at any one time and the *tert*-butyl substituent is often referred as a "*conformational lock*" for the cyclohexane ring. This "locking" has important consequences when other substituents are also present. [In fact the cyclohexane ring is *conformationally biased* into a geometry in which the *t*-Bu group is equatorial rather than being truly conformationally locked since conformational inversion still occurs but the equilibrium is biased strongly to one side]

The free-energy differences (usually referred to as *A*-values) between the two chair conformations (*i.e.* substituent axial or equatorial) have been determined experimentally for a large number of mono-substituted cyclohexanes. Values for some common substituents are shown below:

Substituent (solvent)	Me	Et	i-Pr	t-Bu	Ph	CN	OH (CS_2)	OH (C_6H_{12})	Cl	Br	COOMe	OTs
A-value (kJ/mol)	7.28	~7.3	~9	~21	11.7	0.84	4.3	2.5	2.2	~2.5	5.2	2.1

2.13.2 Disubstituted cyclohexanes Introduction of a second substituent increases the complexity of the problem since there are seven structural and configurational isomers to be considered (the 1,1-, and the *cis* and *trans* isomers of the 1,2-, 1,3- and 1,4-disubstituted cyclohexane). For each of these cases there exist a number of different conformational possibilities.

Substituents of similar size: the dimethylcyclohexanes. There are four dimethyl-cyclohexanes: the 1,1-derivative will not be considered further, but for each of the 1,2-, 1,3- and 1,4-derivatives there is a *cis*- and a *trans*-isomer. In addition there is the possibility of optical isomerism.

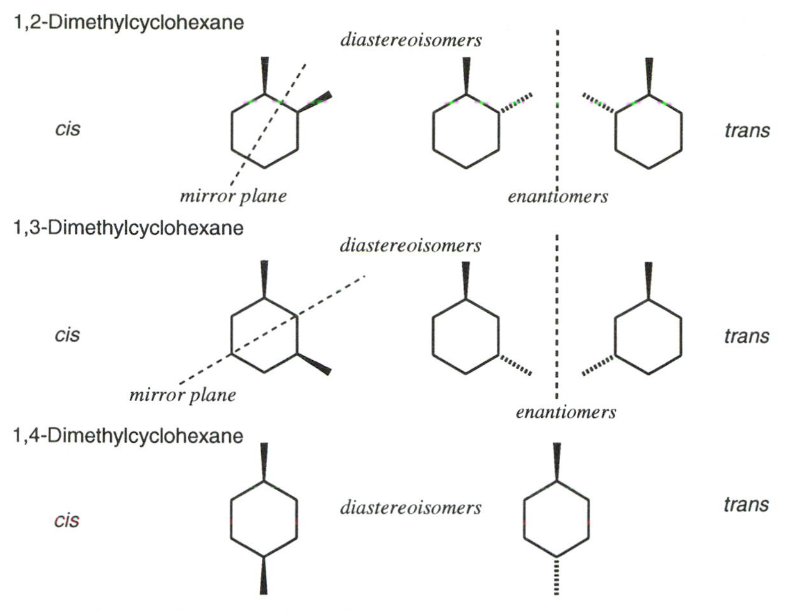

n.b. In flexible structures like the dimethylcyclohexanes, the various [configurational] stereoisomers are most readily identified by consideration of a planar ring geometry.

The relative energies of the configurational isomers have been measured by equilibration studies (*i.e.* determination of the position of the *cis~trans* equilibrium in each case) using a Pd/C catalyst. However the results obtained do not, of course, relate to a flat ring and for each isomer the favoured conformation must also be considered.

1,2-Dimethylcyclohexane. The equilibration studies reveal that the *trans* isomer is more stable by 6.9 kJ mol^{-1}. Either chair conformation of the *cis* isomer must have one methyl group axial and the other equatorial, there being rapid ring inversion between the two equivalent but relatively inverted chair conformers.

Examination of models of the two relatively inverted chair conformations of *cis*-1,2-dimethyl cyclohexane will show that these are related as a pair of enantiomers. However, rapid equilibration between of the two structures results in the lack of optical activity.

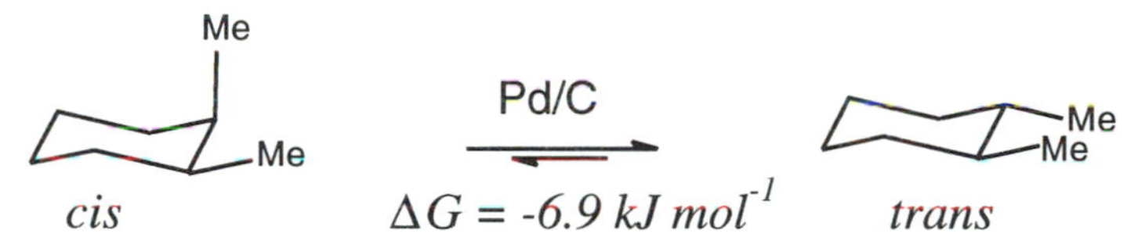

cis $\Delta G = -6.9\ kJ\ mol^{-1}$ *trans*

In the *trans* isomer both substituents can be either diaxial or diequatorial. The diaxial arrangement is destabilised by four gauche interactions (two between each methyl group and the ring skeleton) whereas the diequatorial case has only one extra gauche interaction (between the two methyl substituents). This leads to an energy difference of *ca.* 10 kJ mol^{-1} in favour of the diequatorial (*trans*) conformation resulting in >99% population of the latter at room temperature.

Comparison of the relative energies of the favoured conformations of the *cis* and *trans* isomers predicts that whilst the substituents introduce three additional gauche interactions in the *cis* isomer, there is only one extra gauche interaction in the *trans* stereoisomer. Accordingly the *trans* isomer is calculated to be more stable by $\delta\Delta G$ = *ca.* 6.7 kJ mol^{-1} in good agreement with the experimental data.

As already noted on page 5 the *cis* isomer is a meso-form whereas *trans*-1,2-dimethyl-cyclohexane exists as a pair of enantiomers.

1,4-Dimethylcyclohexane. In this case the equilibration studies reveal that the *trans* isomer is more stable by 6.5 kJ mol^{-1}.

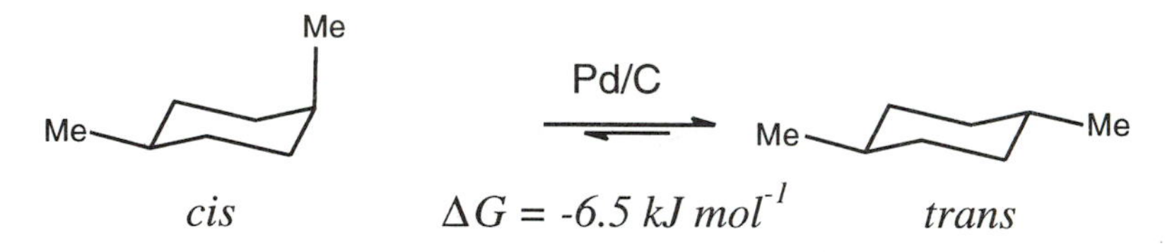

The *cis* isomer must also have one axial and one equatorial methyl group in either chair conformation, whereas in the *trans* isomer both substituents are either diaxial or diequatorial. However, in the latter situation the methyl substituents are sufficiently remote from each other that there is no gauche interaction between them. Once again the *trans* isomer is predicted to be more stable by 6.7 kJ mol^{-1}. Both isomers have a plane of symmetry running through them (the *trans* isomer also has a centre of symmetry) and are therefore achiral.

Contrary to classical stereochemical predictions, *cis*-1,3-dimethyl derivative is more stable than its *trans* isomer (whereas the opposite is true for the 1,2- and 1,4-derivatives). Such a result could not be satisfactorily explained before the development of conformational analysis.

1,3-Dimethylcyclohexane. The *cis* isomer has either diequatorial or diaxial methyl groups. There are no gauche interactions in former conformation whereas there are four in the diaxial case. This latter geometry is further destabilised by a 1,3-diaxial interaction between the methyl groups which results in an overall energy difference of *ca.* 22.6 kJ mol^{-1} between the two conformations.

The *trans* isomer has one axial and one equatorial methyl group in either chair conformation. Inversion of the ring simply converts one enantiomer into itself again. This isomer is potentially resolvable whereas the *cis* isomer has a plane of symmetry (see page 5).

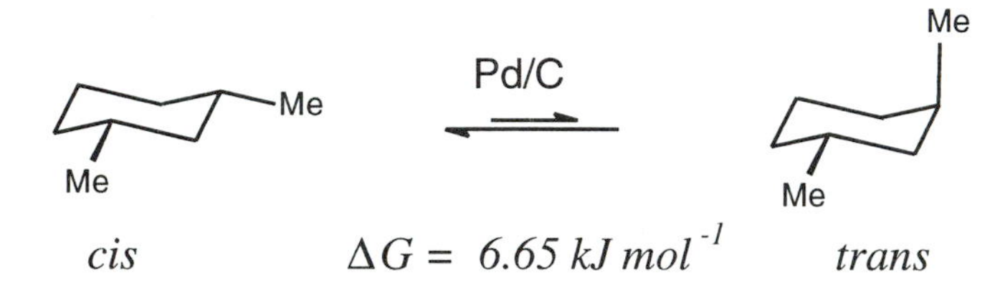

*The effect of larger substituents: the di-*tert*-butylcyclohexanes.* The chair conformation of the *trans*-1,3-isomer must have one *tert*-butyl group equatorial and the other axial thereby destabilising the geometry by *ca.* 21 kJ mol^{-1}. Such a value raises the ground state energy close to that of a boat conformation and consequently a twist-boat in which both the *tert*-butyl groups are located at boat-equatorial sites is preferred.

The steric bulk of the *t*-butyl group forces *trans*-1,3-di-*tert*-butyl cyclohexane into a twist boat conformation.

Other twist-boat forms require one of the *tert*-butyl groups to occupy a boat-axial location.

cis-1,4-Di-*tert*-butylcyclohexane favours a boat geometry in which the substituents are in the "bowsprit" position.

In order to accomodate the bulk of the *tert*-butyl substituents, *cis*-1,4-di-*tert*-butylcyclo-hexane is forced into a boat conformation in which the two bulky substituents occupy the sterically more favourable boat-equatorial sites.

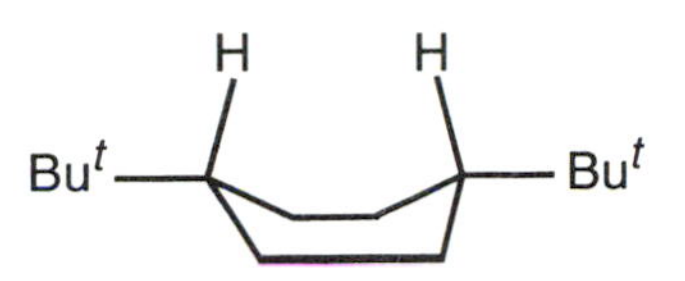

2.13.3 Substituents with large steric bulk differences When the two substituents have significantly different bulk it is the larger group which controls the favoured cyclohexane conformation. For example, in the isomeric 4-*tert*-butylcyclohexanols the *tert*-butyl group demands an equatorial location and thus locks the orientation of the hydroxyl group giving effectively conformationally pure axial (*cis*) and equatorial (*trans*) cyclohexanols:

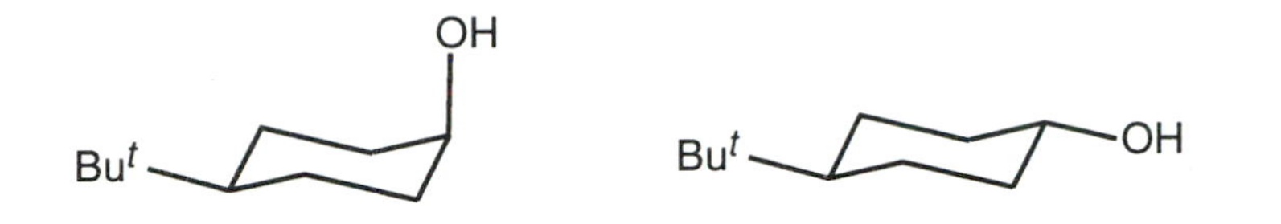

This ability to lock the geometry of a particular substituent on a cyclohexane ring has proved invaluable for the investigation of the relative reactivity of different sites in the molecule as will be discussed later.

2.13.4 Anomalies The presence of polar substituents may introduce other factors which can be more important than the effect of simple steric bulk. For example, *cis*-cyclohexane-1,3-diol prefers a diaxial conformation since this allows intramolecular hydrogen bonding; indeed, infrared studies suggest that this interaction is strong in solution. A 1,3-diaxial geometry for the two substituents is a prerequisite for this to occur (note the formation of another 6-membered ring by the hydrogen bond).

Intramolecular hydrogen bonding can perturb the conformational preference of a substituent.

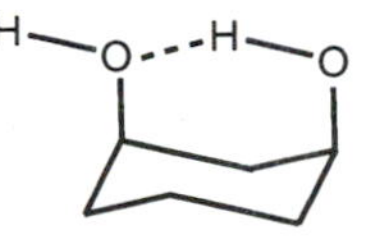

Strong dipole-dipole interactions can also be important. On steric grounds *trans*-1,2-dibromocyclohexane should adopt a diequatorial chair conformation, but dipole moment measurements show that there are substantial amounts of the diaxial form present in solution and that the position of this equilibrium is strongly solvent dependent.

High dielectric constant media disperse the dipoles through solvation and so steric effects dominate (favouring a diequatorial geometry in which the C-Br bonds are 60° apart i.e. a gauche interaction) but in non-polar solvents electronic considerations favour the diaxial conformation despite unfavourable steric interactions since this situation places the dipoles antiparallel so that they cancel each other out.

2.14 Unsaturated systems

2.14.1 Cyclohexanone The introduction of one sp^2-hybridised carbon atom into the cyclohexane ring, as in cyclohexanone, removes one axial and one equatorial hydrogen atom and this molecule adopts a conformation which is only slightly distorted from a normal chair geometry. Cyclohexanone is 12.5 kJ mol^{-1} more strained than cyclohexane; the C–C(=O)–C bond angle is slightly reduced (115°) and the chair is somewhat flattened at the carbonyl end. There is also a twist form which is of relatively low energy (11.4 kJ mol^{-1} above the chair) and chair-chair inversion is much easier than for cyclohexane ($\Delta G^{\ddagger}$ = 16.7 kJ mol^{-1}).

Cyclohexanone adopts a chair conformation:

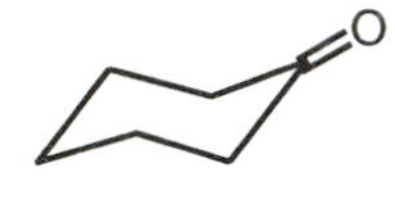

Whilst 2-methylcyclohexanone behaves very like its cyclohexane analogue, axial orientation of ethyl and isopropyl substituents is more favourable because of reduced steric interactions through the presence of the carbonyl substituent. The 2-*tert*-butyl derivative appears to exist predominantly in the twist form.

2-Halocyclohexanones favour an axial conformation (only in low dielectric solvents in the case of the chloro- derivative but in all solvents for the bromo- analogue).

The solvent dependence of the conformational behaviour of 2-halocyclohexanones suggests that dipole-dipole interactions play an important rôle, but for the larger halogens steric interactions are probably also involved.

In the equatorial conformation, the C-X and C=O dipoles lie nearly parallel and repel, whereas in the axial case they are nearly perpendicular and interaction is minimised. It has also been suggested that the anomeric effect (stabilisation through interaction between the C-X σ^*–MO and the π–MO of the carbonyl group) may help to stabilise the axial conformer, particularly for the less electronegative bromo-derivative.

Dipole-dipole interactions may also help to explain the conformational dependence of the carbonyl infra-red stretching frequency in these compounds. An equatorial C–Br bond raises $\nu_{C=O}$ by 20 cm^{-1} whereas an axial halogen substituent has little effect.

2.14.2. Cyclohexenes The presence of a double bond in an alicyclic ring introduces a number of additional constraints. Firstly there is the need to accomodate two *ca.* 120° bond angles within the structure. An additional conformational constraint arises since both the alkene carbon atoms and their immediate neighbours need to be coplanar in order to maintain efficient π-overlap. In order to accomodate this, cyclohexene adopts a half-chair conformation which is inherently flexible. Once again half-chair—half-chair inversion occurs through a "half-boat" geometry (which itself corresponds with an energy minimum in marked contrast with a cyclohexane boat) though this latter is not thought to be the energy minimum intermediate. The activation barrier for such a process (22 kJ mol^{-1} - determined by ^{1}H nmr experiments on cyclohexene-1,1,2,3,4,4-d_6) is significantly lower than that observed for cyclohexane. [Calculations suggest that the half-chair is *ca.* 11 kJ mol^{-1} more stable than the half-boat.]

Cyclohexene adopts a half-chair conformation in which substituents are located either quasiaxial (a') or quasiequatorial (e') as shown in the schematic representation below:

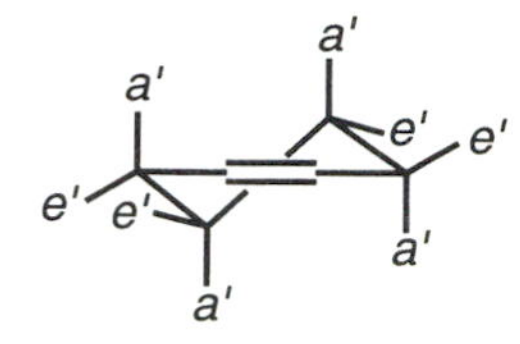

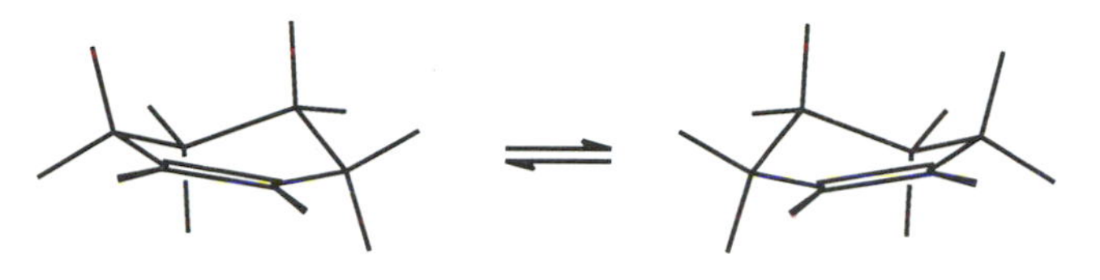

ΔG‡ = 22.2 kJ mol^{-1}

The views to the left provide a more realistic representation of these two half-chair conformations which undergo rapid interconversion at room temperature.

2.14.3 Cyclohexadienes The introduction of additional unsaturation further flattens the ring and lowers the energy barrier for conformational inversion.

Two isomers are known, the 1,3- and 1,4-dienes (the 1,2-derivative in which the double bonds form an allene unit is too strained to be prepared). Whilst the double bonds are conjugated in the 1,3-diene, it is only 4 kJ mol^{-1} more stable than the non-conjugated 1,4-isomer, reflecting the importance of other factors in determining the stabilities of these systems.

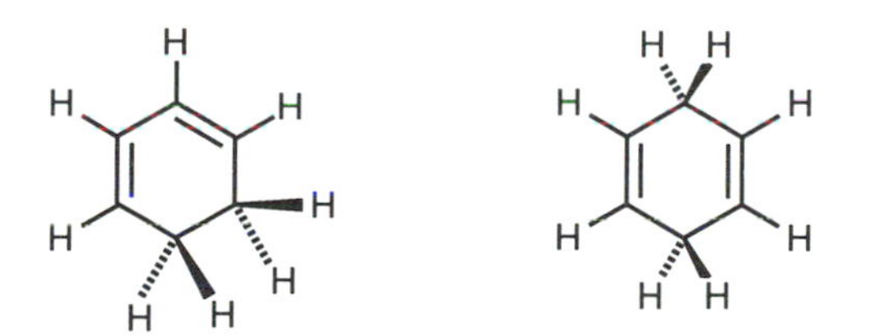

Cyclohexa-1,4-diene (1,4-dihydrobenzene). Cyclohexa-1,4-diene favours a planar energy minimum conformation. This molecule behaves like a book capable of undergoing boat-boat inversion because of the geometric requirements of the two alkene moieties.

Cyclohexa-1,4-diene adopts a planar energy minimum geometry but undergoes wide-amplitude boat-boat vibrations at room temperature.

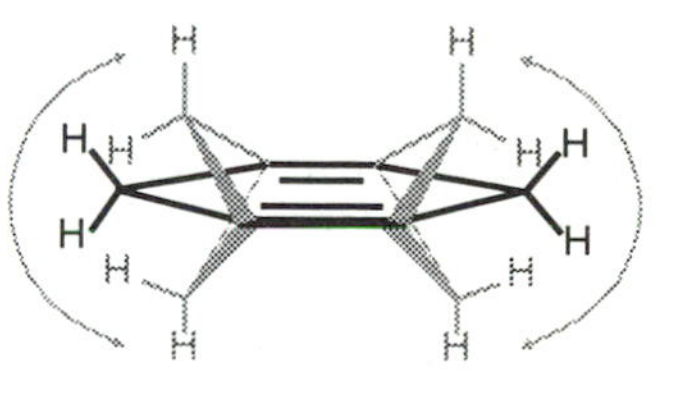

When the ring is flat, torsional interactions between the methylene and alkene C-H groups are minimised but angle strain is high. A boat geometry minimises angle strain at the expense of the torsional interactions. The combination of these two competing effects results in the molecule occupying a very shallow potential energy well with a

Torsional interactions about a cyclohexa-1,4-diene ring:

(a) in a planar geometry;

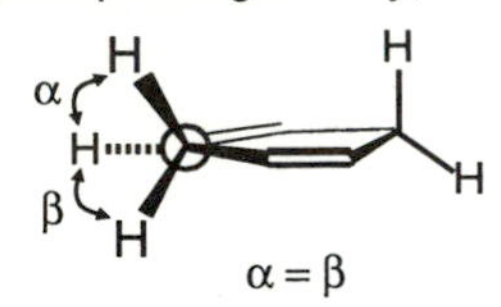

(b) in a puckered geometry.

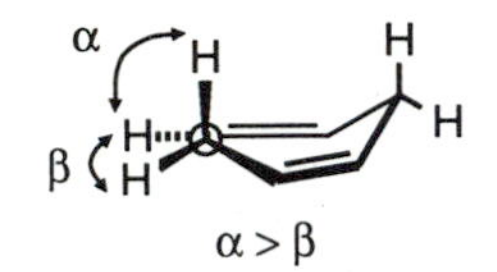

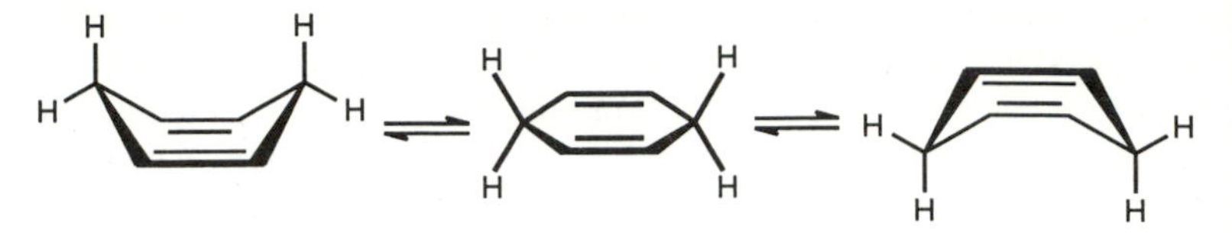

single minimum corresponding to a planar geometry and this is what is observed by X-ray structural studies at low temperatures. However, at room temperature this molecule undergoes wide-amplitude vibrations corresponding to boat-boat inversion as has been demonstrated by far-infra-red and variable temperature nmr experiments

Cyclohexa-1,3-diene. The 1,3-diene isomer contains a butadiene moiety which is capable of twist. This allows the molecule to adopt a relatively unstrained pseudo half-chair geometry which can flip into its mirror image (no boat conformation is accessible in this case).

2.14.4 Other Cycloalkenes Cycloalkenes can be formed for all ring sizes. The presence of a double bond in a ring introduces the possibility of *cis/trans* isomerisation. However a short methylene chain cannot be stretched across the diagonally opposite ends of an ethene fragment. Consequently *trans*-cyclo-octene is the smallest *trans*-cycloalkene that can be satisfactorily prepared, though as will be seen later even this molecule is not very stable. None-the-less there is evidence for the transient existence of *trans*-cyclohexene, and *trans*-cycloheptene is moderately stable at -78°C.

A number of consequences follow from the introduction of a double bond into an alicyclic ring:

- a carbon-carbon double bond is shorter than a single bond ($r_{C=C}$ 136 pm; r_{C-C} 154 pm);
- the presence of a C=C unit introduces two larger bond angles (C-C=C = 120°);
- each alkene unit removes two C-H substituents.

Two naturally occurring cyclopropenes:

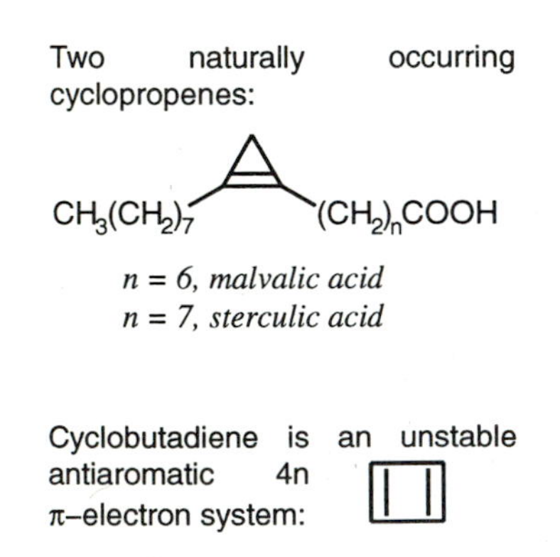

n = 6, malvalic acid
n = 7, sterculic acid

Cyclobutadiene is an unstable antiaromatic 4n π–electron system:

Although introduction of a carbon-carbon double bond into a three-membered ring relieves CH-CH eclipsing interactions, the bond angle requirements of two sp^2 hybridised carbon atoms are considerable and cyclopropene is very strained (218 kJ mol^{-1}). None-the-less two cyclopropene derivatives, sterculic and malvalic acid, occur naturally in seed oils.

The strain in cyclobutene is somewhat less (121 kJ mol^{-1}, only ~13 kJ mol^{-1} more than cyclobutane itself) despite a C=C–C bond angle of 94° which is compensated for by reduced CH-CH torsional interactions. Cyclobutadiene is extremely unstable not just as a result of its considerable bond-angle strain but also because of its "antiaromatic" character (4n cyclic delocalised π-electrons)

Whilst the increased bond angle presents problems for small rings it can be beneficial to the stability of medium ring structures. In addition unsaturation removes two C-H bonds thereby reducing transannular interactions in medium rings.

Heats of hydrogenation provide a convenient measure of the relative stabilities of different cycloalkenes. Comparison of the values for the various *cis*-cycloalkenes with that for cyclohexene (ΔH_n - ΔH_6) reveals the relative instability of the small rings (C_3 and C_4), but the negative values measured for C_5, and C_7 to C_{10} reflect the introduction of torsional strain and non-bonded interactions resulting from the addition of two hydrogen atoms across the double bond.

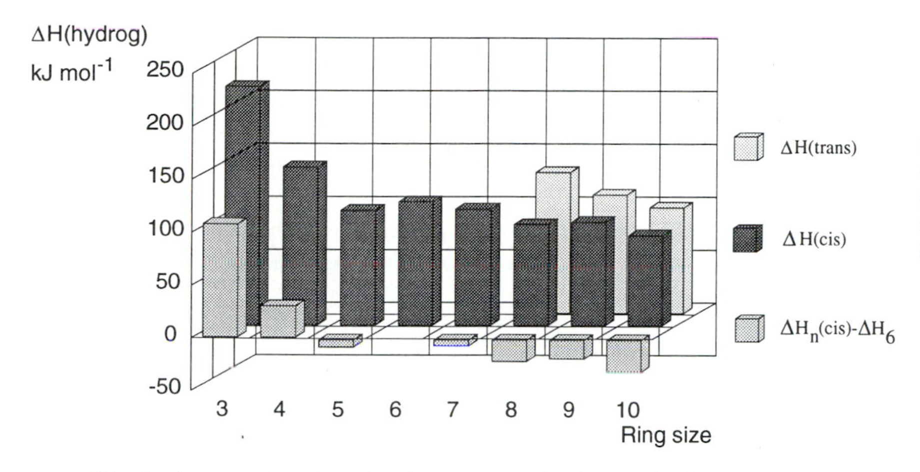

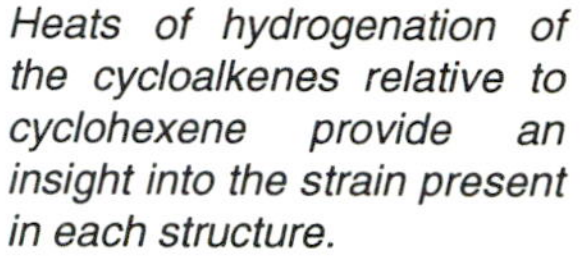

Heats of hydrogenation of the cycloalkenes relative to cyclohexene provide an insight into the strain present in each structure.

The hydrogenation data for the *trans*-cycloalkenes indicates their relatively strained nature, and that strain is relieved as the ring size increases. Indeed in cycloundecene and cyclododecene the *cis* and *trans* isomers are of comparable energy as is shown by the equilibrium constant data for acid-catalysed *cis/trans* equilibration (in naphthalene-2-sulfonic acid at 373K). In even larger rings there is minimal strain and the *trans*-isomer is significantly favoured, as is normally the case in acyclic systems.

Acid-catalysed *cis/trans*-equilibration of cycloalkenes.				
Ring size	9	10	11	12
K(cis/trans)	232	122	0.41	0.52

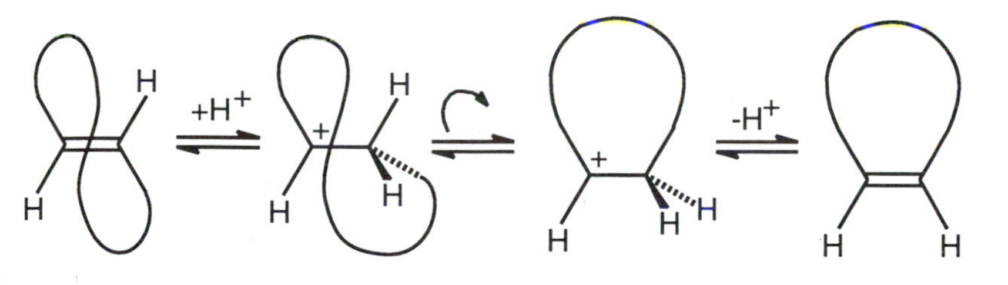

trans-Cyclo-octene is a dissymmetric molecule and therefore exists as a pair of enantiomers:

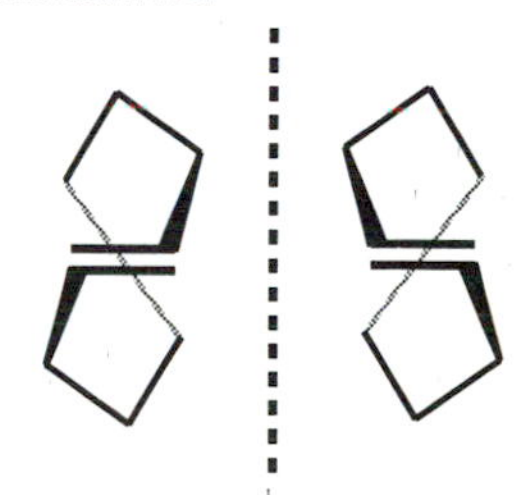

2.14.5 Chirality in trans-cycloalkenes Another important feature of the chemistry of the *trans*-cycloalkenes arises from the dissymmetry present in these molecules which results from twisting of the methylene chain linking the ends of the double bond, the preferred conformation of *trans*-cyclo-octene being crown-like.

Such molecules racemise by rotation of the methylene chain but the very compressed nature of *trans*-cyclo-octene makes this comparatively slow. Such a process does not lead directly to the crown-enantiomer but procedes via a distorted chair conformation which is calculated to be *ca.* 13 kJ mol^{-1} less stable than the crown:

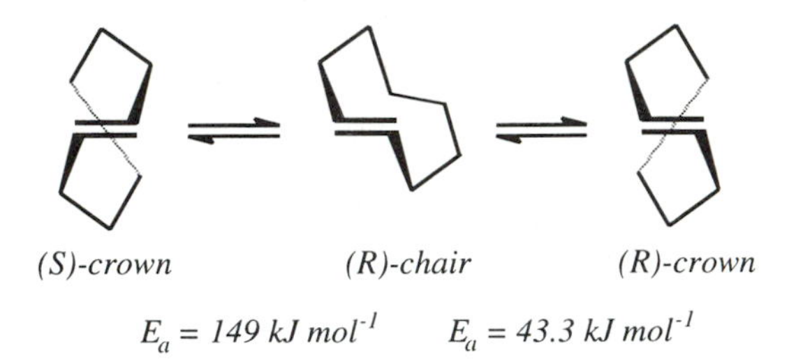

As the ring size is increased there is a marked reduction in the activation energy for racemisation reflecting increased flexibility of the longer methylene chains:

Racemisation of *trans*-cycloalkenes			
Ring size	8	9	10
$\Delta G^{\ddagger}$ (kJ/mol)	149	84	45

Cyclo-octyne is the smallest isolable cycloalkyne:

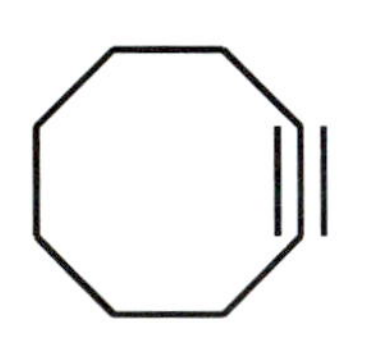

2.14.6 Cycloalkynes Whilst there is good evidence for the existance of cycloheptyne, cyclohexyne and even cyclopentyne as transient reaction intermediates, cyclo-octyne is the smallest moderately stable cycloalkyne. A complication which arises in these structures is the possibility of alkyne-allene (cycloallene = cycloalka-1,2-diene) isomerisation particularly under basic conditions. Normally the allene is more stable than the corresponding alkyne but the enthalpy difference is not large and, in some medium rings, particularly cyclononyne, other factors can reverse the normal trend. For example, base-catalysed equilibration leads to allene/alkyne ratios:

Base-catalysed allene/alkyne equilibration			
Ring size	9	10	11
Allene/alkyne ratio	20	1	0.1

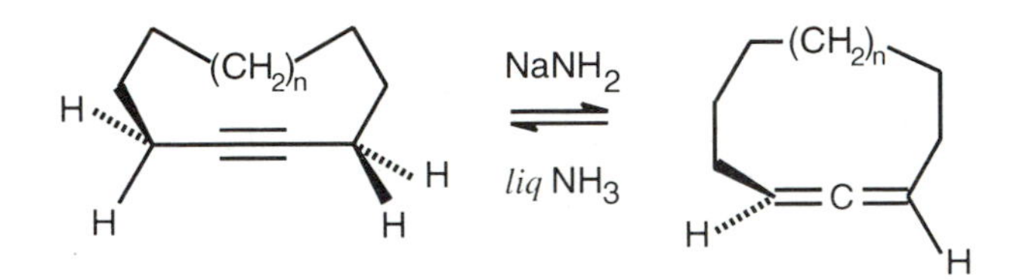

2.15 Effect of strain on reactivity

Reactions which are likely to be influenced by ring-size are those which involve a change in the co-ordination number at the reaction centre *e.g.* through a change of hybridisation from sp^3 to sp^2 or *vice-versa*. There are two classes of reaction to be considered:

(i) Equilibrium-controlled processes

Here the outcome of the reaction (the position of the equilibrium) will depend solely on the relative thermodynamic stabilities of the starting materials and the products;

(ii) Kinetically controlled processes

In this case the important factor is the activation energy (*i.e.* the energy difference between the reactants and the rate-determining transition state).

There is a concept known as "*I*-strain". This refers to the strain changes brought about by a change of hybridisation which produces concomitant changes in angle, torsional and transannular strain all of which contribute to the I-strain.

2.15.1 Properties and reactions of cycloalkanones

2.15.1.1 Infra-red spectroscopy. The infra-red stretching frequency of a carbonyl group is very sensitive to bond angle. For acetone ν_{max}(C=O) is 1718 cm^{-1} but as the C–C(=O)–C bond angle increases ν_{max} falls though there is also considerable sensitivity to the nature of the substituents present. The carbonyl stretching frequencies of the cycloalkanones show a marked variation with ring size. The values for the medium rings indicate that in each case the C–(CO)–C bond angle is significantly larger than that in cyclohexanone, consistent with X-ray structural studies on these ketones.

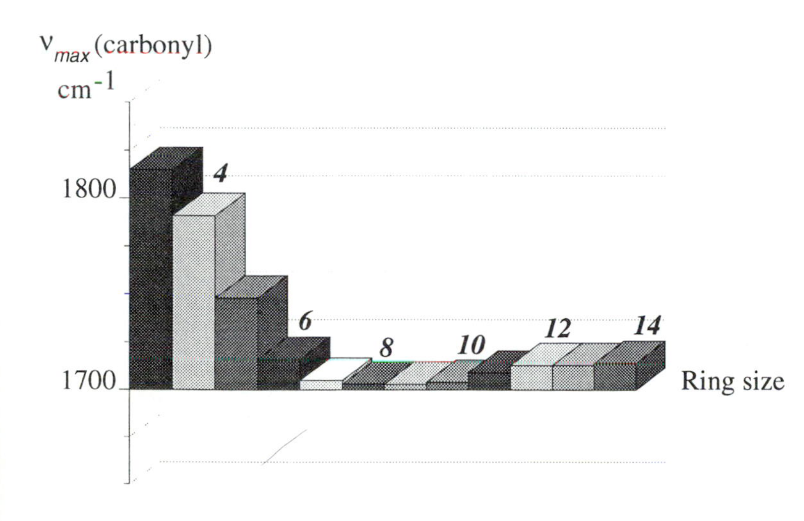

Infra-red stretching frequencies for carbonyl groups in cycloalkanones.

2.15.2 Equilibrium-controlled reactions Reactions of ketones normally involve addition to the carbonyl group leading to a change of hybridisation of the carbonyl carbon from sp^2 to sp^3.

2.15.2.1 Cyanohydrin formation. Carbonyl compounds react with KCN (in the presence of an acid catalyst) to form cyanohydrins. This process is an equilibrium and though efficient for aldehydes, is not normally favourable for ketones because of steric crowding arising from conversion of a trigonal (sp^2) carbonyl carbon atom into a tetra-substituted (sp^3) centre.

Equilibrium constants (K) for cyanohydrin formation:

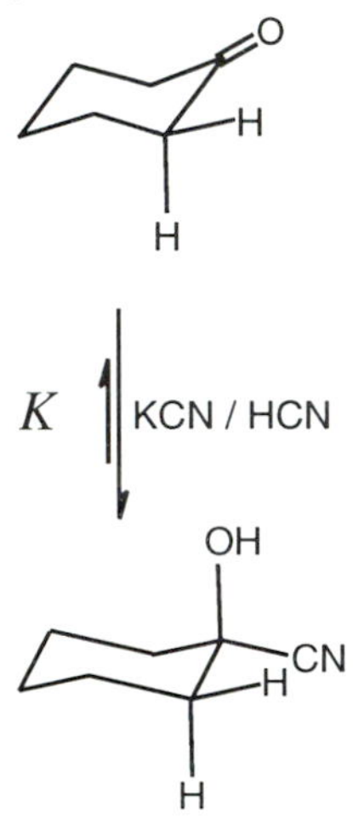

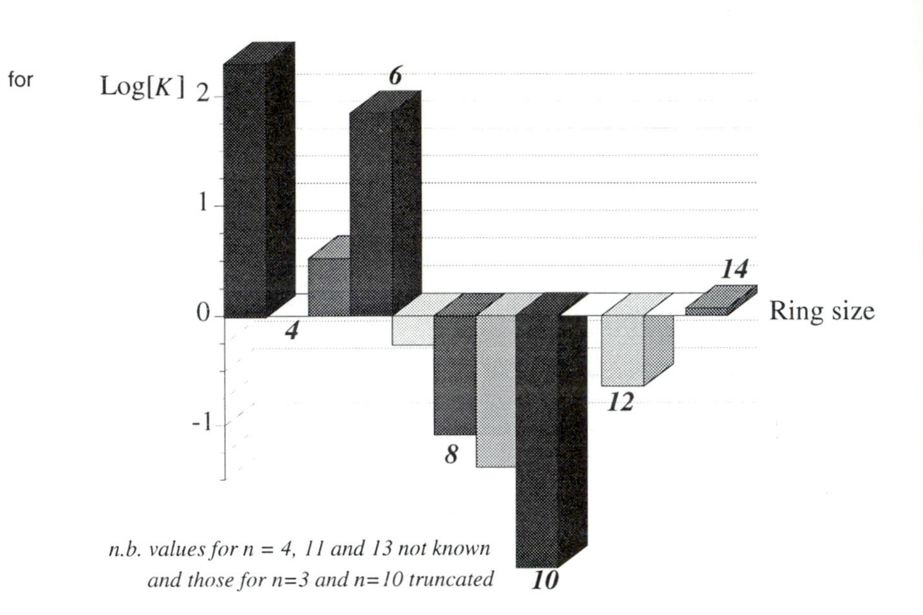

For cyclohexanone this equilibrium lies in favour of the cyanohydrin reflecting the relief of angle strain and eclipsing interactions between the carbonyl bond and neighbouring equatorial C-H bonds present in the ketone.

For cyclopropanone K is even larger reflecting the problems of accommodating an sp^2-hybridised carbon atom in a very small ring (relief of angle strain dominates, favouring cyanohydrin formation). The importance of this effect is further illustrated by the fact that cyclopropanone is completely hydrated in water.

For cyclopentanone, cyanohydrin formation is much less favourable since eclipsing interactions are important in the product which reduce its stability relative to the ketone.

As the ring becomes larger, cyanohydrin formation becomes rather unfavourable. In cycloheptanone eclipsing interactions again become significant in the cyanohydrin and in the medium rings the presence of an sp^2 centre reduces the number of C-H bonds to be accommodated thereby providing some relief of transannular non-bonded repulsions and torsional interactions. In addition, the saturated structures are beset with bond-angle deformations (widening) which are also reduced in the ketone.

2.15.3 Kinetically controlled reactions

2.15.3.1 Solvolysis of tosylates. Since this process occurs by an S_N1 mechanism, reaction rates give an insight into the advantages of having an sp^2 carbon centre developing in the ring as the C-OTs bond breaks.

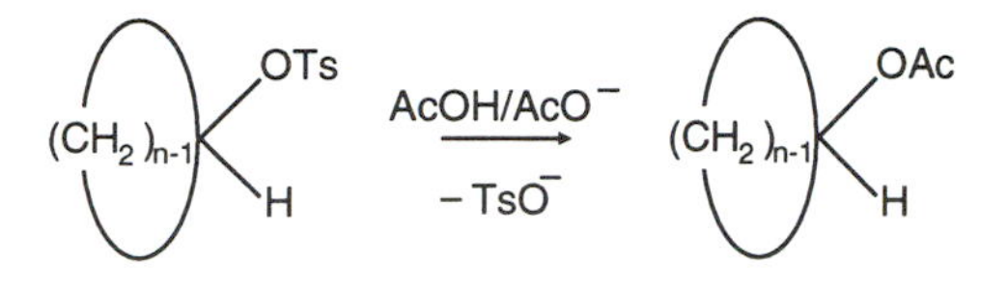

The rate of acetolysis of cyclohexyl tosylate and the tosylates of large cycloalkanes ($n \geq 14$) is typical of normal acyclic secondary derivatives. Indeed, log $k_{(acetolysis)}$ and ν_{max} (C=O) show a reasonable inverse relationship for ring sizes in the range $n = 6 - 15$.

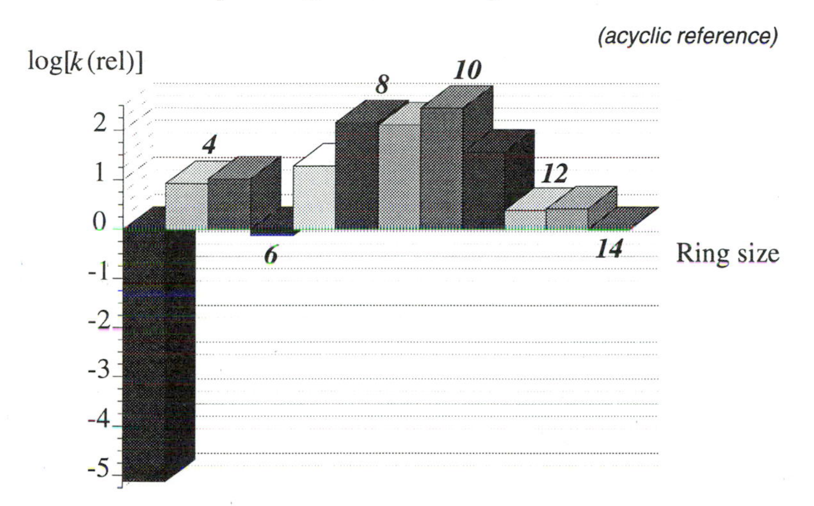

Relative rates (acyclic reference) of acetolysis of cycloalkyl tosylates.

Reaction of the cyclopropyl ester is very much less favourable. In this case the C-OTs bond is relatively strong (reflecting high "s"-character at carbon) and ionisation is hindered by angle strain caused by a developing sp^2 centre in a three-membered ring.

For all other ring sizes (*i.e.* n = 4, 5, 7-13) rate enhancements are observed. The process is particularly favourable for the medium ring esters where relief of transannular interactions and of angle strain, already present through C-C-C bond angle widening, can even provide assistance to the reaction.

The relatively high reactivity of the cyclobutyl ester is perhaps surprising but in this case special factors operate involving the formation of an unusually stable carbocation intermediate showing non-classical behaviour. Its structure may be best represented as a resonance between a cyclobutyl, a cyclopropylcarbinyl and a homoallyl cation:

Reactions of cyclopropyl-carbinyl, cyclobutyl and homoallyl derivatives often appear to proceed *via* a common intermediate which has the character of a resonance structure lying somewhere between those of each of the initially formed cations.

Capture by nucleophiles, *e.g.* solvent, leads to a mixture of products derived from each of the possible starting materials.

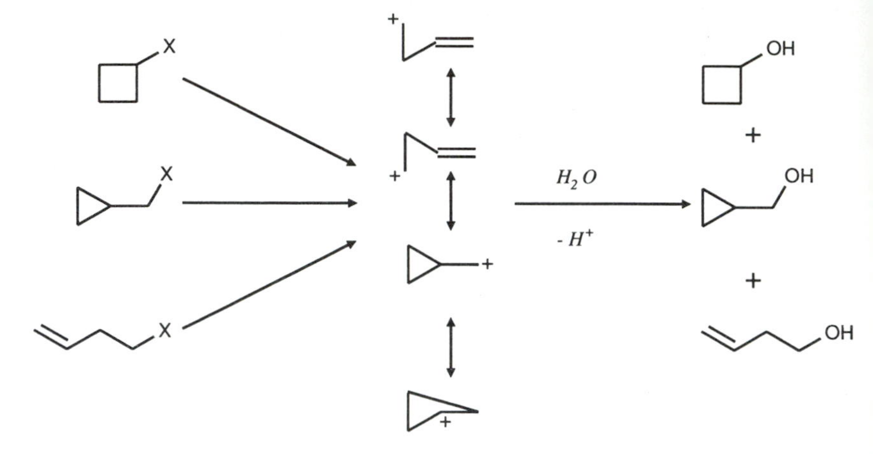

2.15.3.2 Borohydride reduction of cycloalkanones. In this process the carbonyl group is being reduced to a secondary alcohol thereby converting an sp^2 centre to sp^3 hybridisation. The data suggest that, relative to an open-chain system, C_4-C_6 ketones prefer the presence of an sp^3 rather than an sp^2 carbon atom in the ring, whereas the C_8-C_{12} ketones prefer the presence of an sp^2 centre.

Relative rates (acyclic reference) of borohydride reduction of cycloalkanones.

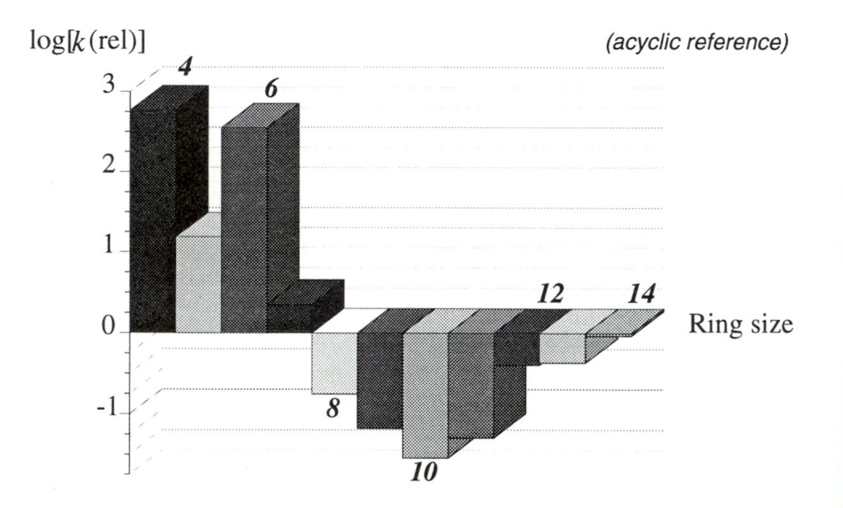

Borohydride reduction of ketones is a kinetically controlled process which is effectively the reverse of the tosylate solvolysis and the two sets of rate data are in general mirror images of each other (bearing in mind the special nature of the solvolysis of cyclobutyl derivatives). This is not surprising since the two reactions involve complementary hybridisation changes. The inverse correlation also suggests complementary changes in molecular compression in the two transition states.

Indeed there is a linear relationship between log $k(BH_4^-$ reduction) and log K (cyanohydrin formation) for alicyclic ketones over the range $n = 5 - 17$ reflecting the similar nature of reduction transition state and the stability of a tetrahedral centre in such rings relative to the presence of a carbonyl group.

2.15.3.3 Formation of a cycloalkyl free-radical centre. The rates of fragmentation of a series of azocycloalkyl nitriles shows a marked dependence on ring size. This process is particularly fast for medium-ring derivatives, again reflecting relief of *I*-strain upon the introduction of an sp^2-site into the cycloalkyl structure.

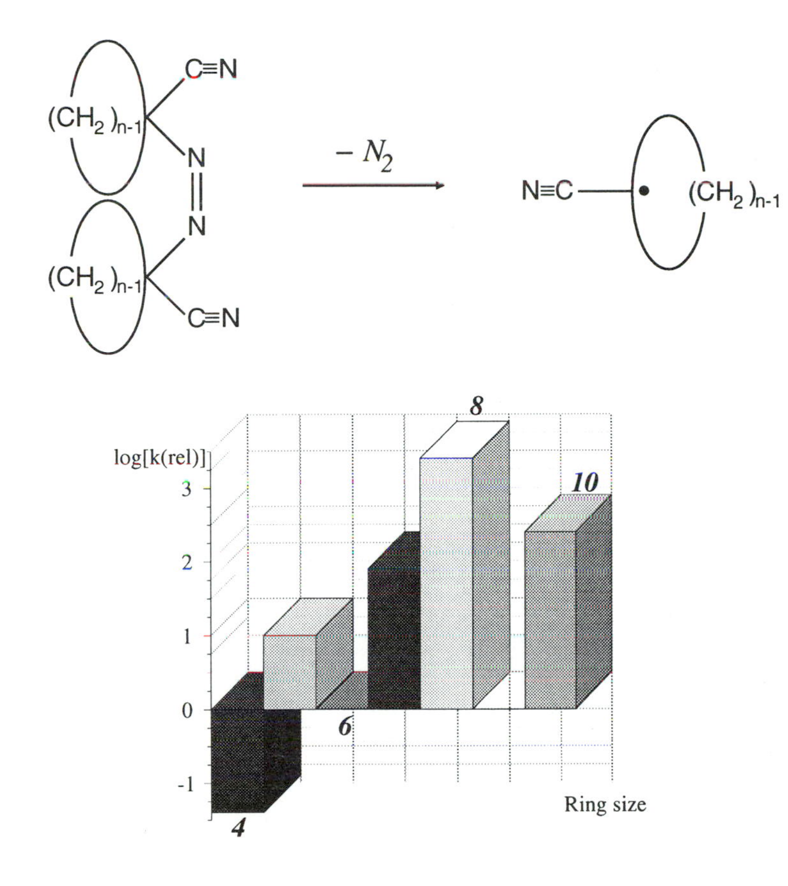

Relative rates of decomposition of azocycloalkyl nitriles.

3: Ring synthesis

Approaches to the synthesis of ring structures obviously provide a key component in the repertoire of synthetic techniques. Different ring sizes impose different constraints and problems requiring a variety of solutions.

3.1 Principles controlling ring-closure reactions

The rate of a reaction depends on the activation energy of the process (strictly $\Delta G^{\ddagger}$); this is made up of two contributions, one enthalpic and the other entropic.

In a kinetically controlled reaction:

$$\text{Rate constant } k = A\, e^{-\frac{\Delta G^{\ddagger}}{kT}}$$

$$\text{where } \Delta G^{\ddagger} = \Delta H^{\ddagger} - T\Delta S^{\ddagger}$$

The enthalpy term ($\Delta H^{\ddagger}$) reflects not only bond making and bond breaking energies but also bond-bond repulsions plus angle and other strain factors present as the ring forms.

A key factor determining the entropy of activation ($\Delta S^{\ddagger}$) is the loss of rotational freedom as the chain precursor is converted into a ring. The longer the chain, the more rotational freedom that is lost. As the chain length increases, $\Delta S^{\ddagger}$ decreases (becomes more negative) because there is statistically less chance of the ends meeting as the chain lengthens, so the entropy term becomes less favourable (*n.b.* a negative $\Delta S^{\ddagger}$ value reflects the increased order of the structure resulting from ring formation). Solvation changes can also significantly affect the entropy term.

Ring synthesis can involve the formation of one:

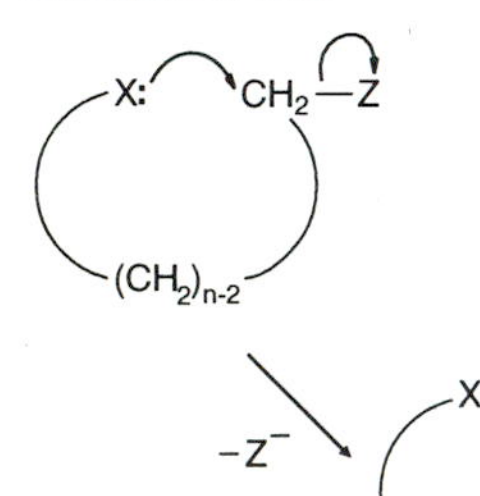

or two bonds at a time:

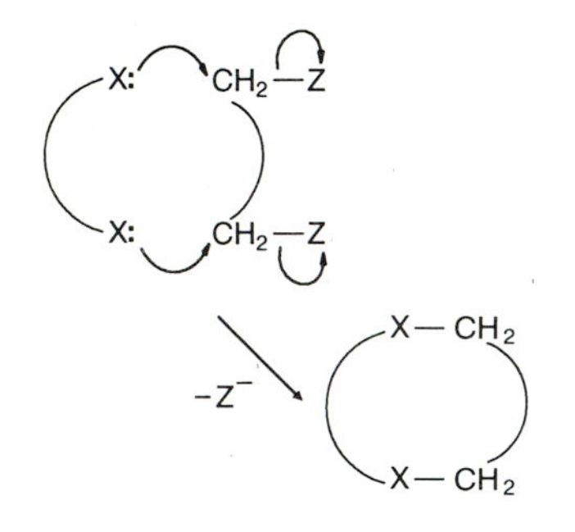

3.2 Kinetic vs. thermodynamic control in ring-forming processes

Most ring-forming reactions fall into one of two basic classes, *i.e.* reactions in which either (a) one bond, or (b) two bonds are formed at one time. Another factor which plays a key rôle in the success of ring-forming processes is whether they are equilibria or irreversible (kinetically controlled) reactions. In general, only 5 and 6-membered rings are formed under equilibrium-controlled conditions because in other cases the cyclic structure is less stable than the open-chain form. However, under kinetic control the energy cost involved in forming a strained ring is overcome by other more favourable energetic factors in the reaction.

For small rings, particularly cyclopropane, angle strain dramatically increases as the ring is formed and torsional (eclipsing) interactions are important in the transition state leading to the planar ring. In practice three-membered rings are relatively easy to form because of the very favourable entropy factor reflecting the close proximity of the sites being linked. Four-membered rings are less readily synthesised because the entropy term is less favourable (the reaction sites are further apart) whilst the strain problem remains significant.

For normal rings the enthalpic contribution is favourable and the entropic term has not become too bad with the result that these rings are readily prepared by a wide variety of reactions.

However, for the medium rings neither enthalpic nor entropic factors favour cyclisation. Consequently such structures have proved to be a synthetic challenge and special techniques such as the use of high dilution conditions are required to encourage ring-formation. Since the rate of cyclisation (*i.e.* intramolecular reaction) k_{intra} α [substrate] whereas that for intermolecular reaction (*i.e.* polymerisation) k_{inter} α $[substrate]^2$, high dilution significantly promotes ring formation.

High dilution promotes ring formation because of the greatly reduced chance of a bimolecular (polymerisation) reaction relative to cyclisation.

For large rings there is no enthalpic penalty arising from cyclisation but the entropy factor is normally very unfavourable since the chances of the two ends of a long chain meeting are minimal. However, under some circumstances preorganisation of the chain around for example, a metal-ion template can overcome such problems. Thus the formation of large macrocyclic crown ethers such as 18-crown-6 (an 18-membered ring containing six oxygen atoms) occurs readily in the presence of K^+ because the polyether chain wraps around the metal ion thereby bringing the chain termini into close proximity.

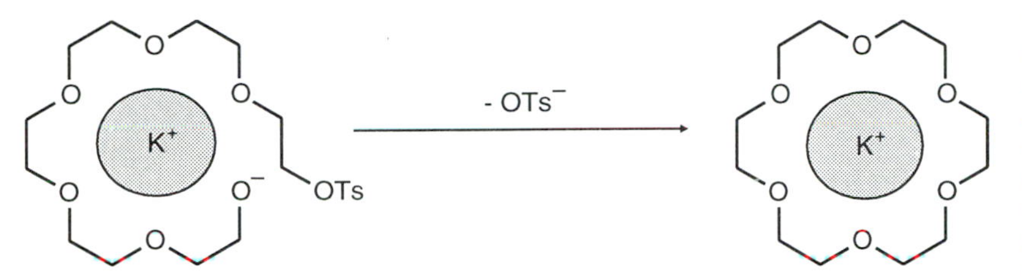

18-Crown-6 has a central cavity ideally suited for the complexation of K^+. When the crown is forming, complexation with this ion brings the reacting sites into close proximity, thereby helping to overcome the large entropic penalty associated with formation of such a large ring.

3.3 Other factors influencing the efficiency of ring closure processes

Many reactions require tightly defined trajectories for attack on the reaction site in order to achieve the optimum geometry required for reaction via a minimum energy transition state. In cyclisation processes the nature of the chain linking the reacting centres often imposes severe constraints on the trajectories available for an intramolecular reaction and this frequently accounts for the failure of particular reaction types.

3.3.1 Stereoelectronic control in cyclisation reactions Frontier Molecular Orbital (FMO) Theory regards chemical reactions as resulting from interaction of the highest occupied molecular orbital (HOMO) of one component and the lowest unoccupied/vacant molecular orbital (LUMO) of the other.

3.3.1.1 The S_N2 reaction. For example, an S_N2 reaction involving reaction of a nucleophile with an alkyl halide arises from interaction of the molecular orbital containing the lone pair on the nucleophile (the HOMO) with the lowest energy antibonding molecular orbital (the LUMO) of the substrate. This latter will be the σ* MO for the carbon-leaving group bond (the weakest bond in the alkyl halide) which lies along the C-X bond axis but with highest probability on the opposite

FMO view of an S_N2 process.

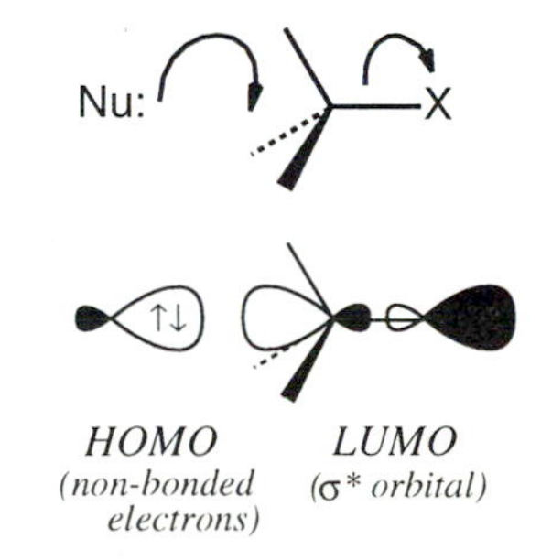

face of the carbon atom from that containing the leaving group X. Consequently the nucleophile, the reaction centre and the leaving group must be collinear in order to optimise the HOMO/LUMO interaction.

A number of elegant experiments have demonstrated this stereoelectronic requirement for the S_N2 process. Intramolecular methyl transfer in the sulfonate ester shown below fails whereas cyclisation of the corresponding iodomethyl derivative is observed.

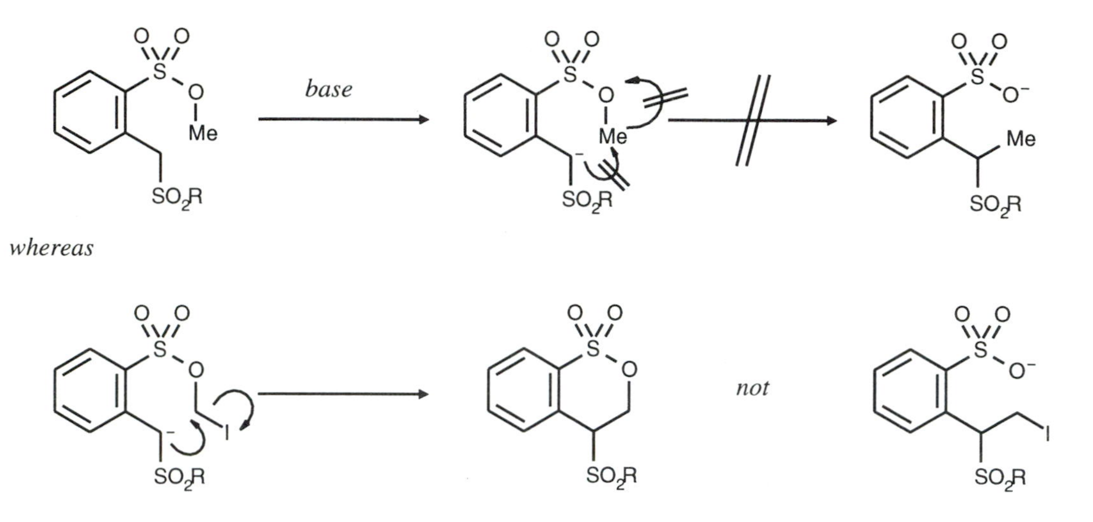

The intramolecular alkyl-group transfer process involves a 6–membered ring transition state which cannot accommodate the linear geometry for the nucleophile, reaction centre and leaving group required by an S_N2 process, whereas the iodide-displacement S_N2 transition state involves a leaving group external to this ring (an *exo*-cyclisation; see 3.3.2), a situation which imposes no such constraints.

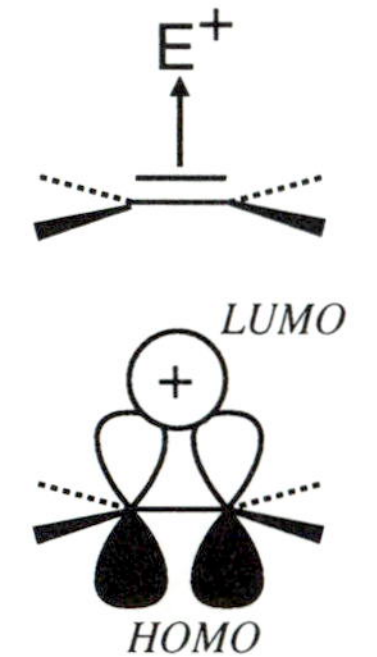

3.3.1.2 Addition to unsaturated groups. In this case there are several possibilities, the simplest of which is attack of an electrophile such as H^+ on an alkene. Clearly the proton must provide the LUMO for the bonding interaction, and therefore the HOMO must be the π MO of the alkene. This has maximum charge density perpendicular to, and in the middle of the alkene plane and thus the initial trajectory of attack should be on the centre of the π-bond. Such considerations are reflected in the fact that carbocation cyclisations to give 6-membered rings occur much more readily those leading to cyclopentyl derivatives:

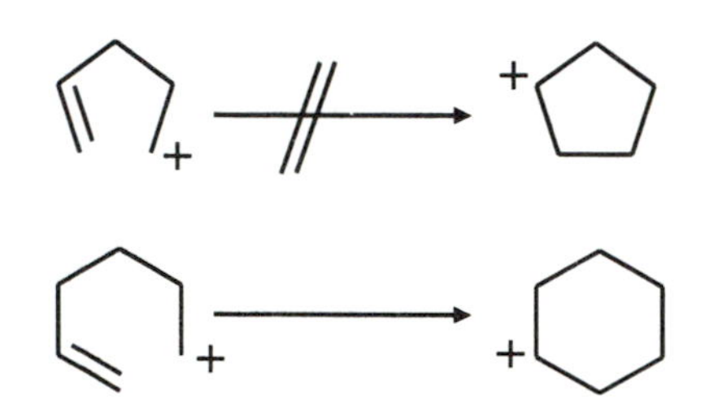

Nucleophilic addition to a simple alkene is not normally observed but occurs readily for a carbonyl group. This process requires interaction of the HOMO, the nucleophile lone pair, with the LUMO of the carbonyl group, in this case the π^* MO. Such a process must clearly occur at one end of and perpendicular to the double bond plane since the π^* MO has a node between the two component atoms.

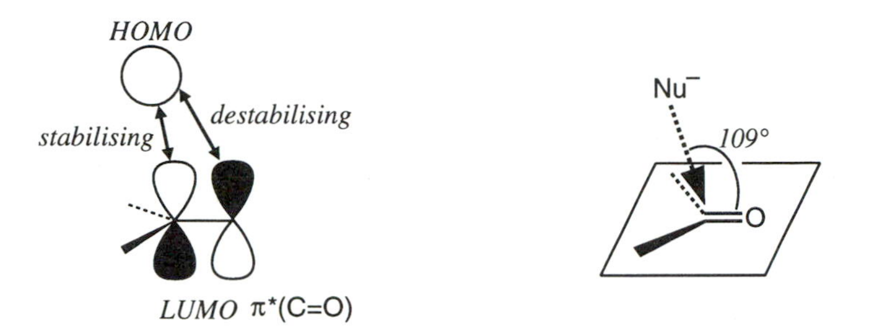

In practice the optimum angle of addition to C=O has been found to be 109°. This tilting away from the carbonyl group is thought to occur in order to avoid destabilising overlap with the component of π^* residing on the oxygen atom and to reduce a destabilising interaction with the HOMO of the substrate (the π MO). Similar considerations should apply to activated, *i.e.* electron-deficient alkenes.

Radical cyclisations involving addition to an alkene are also synthetically important. These involve interaction between a SOMO (a singly occupied MO *i.e.* the radical centre) and the LUMO of the π bond (*i.e.* the π^* MO) and should therefore mimic the nucleophilic addition process. One consequence of this is that the hex-5-enyl radical cyclises exclusively to give a primary cyclopentylmethyl radical rather than the more stable 6-membered ring alternative reflecting problems in achieving a satisfactory six-ring cyclisation trajectory in the latter case. Similarly the but–3–enyl radical prefers cyclisation to a 3-membered ring:

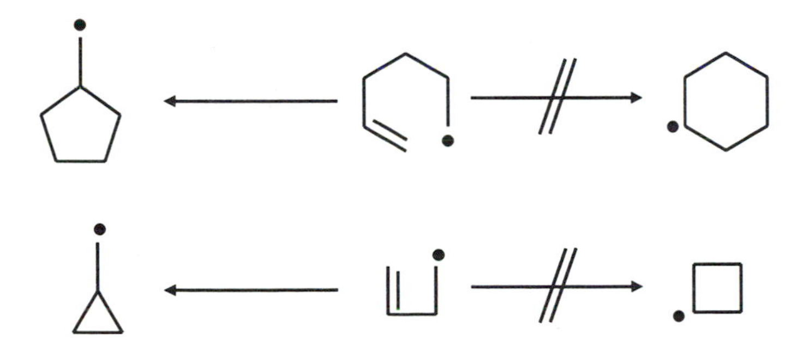

3.3.2 Some rules for ring-closure A set of guidelines have been developed (known as Baldwin's rules) which predict whether or not a particular ring closure should occur to give small and normal rings. The nomenclature used to classify nucleophilic cyclisation processes has the general form:

N–*exo*/*endo*–*dig*/*trig*/*tet*

where:

- N is the ring size of the transition state;
- *exo* or *endo* refer to the relationship between the nucleophile and the bond being broken, *endo* if the two atoms involved in the bond being broken are part of the ring being formed or *exo* if the bond being broken is external to the resulting ring;
- *dig*, *trig* or *tet* indicates the nature of the reaction centre, *i.e.* tetrahedral (*tet*), trigonal (*trig*) or linear (digonal, *dig*).

exo *endo*

tet

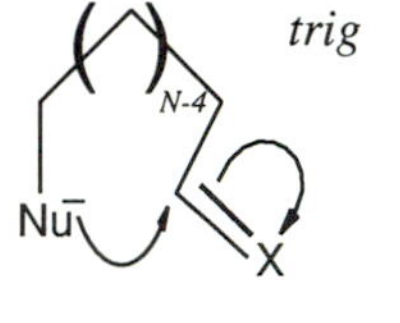

trig

dig

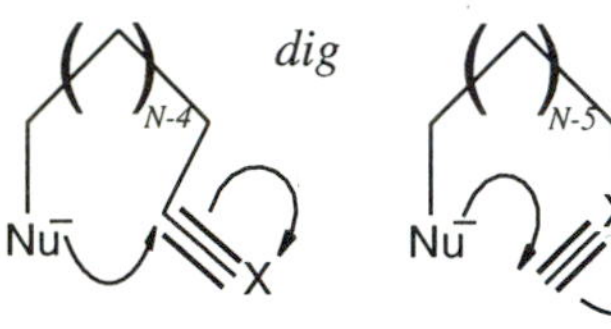

The table below indicates whether particular processes are favoured or unlikely.

Favoured	Disfavoured
3 to 7 *exo-tet*	5 and 6 *endo-tet*
3 to 7 *exo-trig*	3 to 5 *endo-trig*
6 and 7 *endo-trig*	3 and 4 *exo-dig*
5 to 7 *exo-dig*	
3 to 7 *endo-dig*	

The competing intramolecular S_N2 reactions described above are therefore a disfavoured 6-*endo-tet* process which fails, and a 6-*exo-tet* reaction which succeeds, both being in accord with prediction.

Examples of nucleophilic reactions involving trigonal centres are conveniently illustrated by the preferred cyclisation pathways for the two enolates shown below:

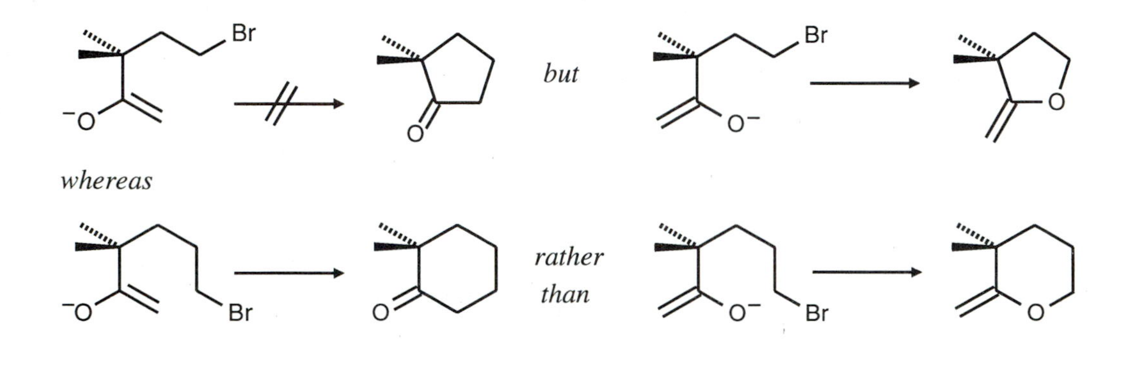

The *endo-trig* reaction involving the enolate carbon readily occurs to give a 6-membered ring but fails to give a cyclopentyl product, cyclisation through the oxygen atom being favoured in this latter case. This result can be readily understood in orbital terms since though the pentenyl chain can adopt an almost planar C-cyclisation transition state in which the C-Br bond lies in the alkene plane, it cannot twist to bring the back of the C-Br bond down onto the alkene π-cloud as required for interaction with the enolate π MO. However, the enolate oxygen atom has lone pairs both perpendicular to and in the plane of the 5-ring transition state and can therefore readily react. No such problems arise for the more flexible 6-membered ring transition state where the normally favoured C-cyclisation is the preferred reaction.

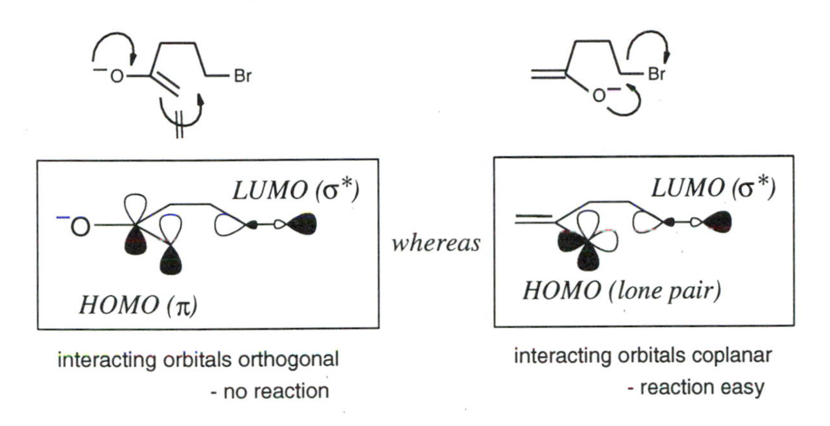

3.4 Making one bond at a time

The majority of ring syntheses occur by bond-formation between two remote sites in a chain.

3.4.1 The Ruzicka ring synthesis The pyrolysis of calcium, thorium or barium salts of α,ω-dicarboxylic acids is an historically important route to cyclic ketones having $n \geq 5$ and provided a valuable method for preparing large-ring ketones which are important to the perfumery industry. Good yields are obtained for cyclopentanone and cyclohexanone and the reaction is also useful for ketones in the range $13 \leq n \leq 19$ but fails for ring sizes between 9 and 12.

The Ruzicka reaction is useful for the preparation of large-ring ketones such as civetone (below) which are valuable perfumes:

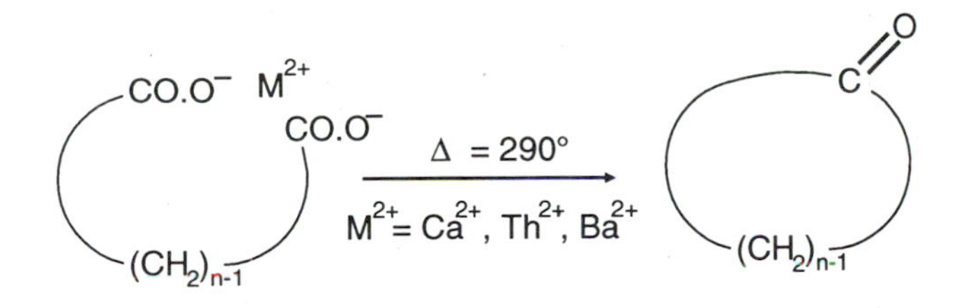

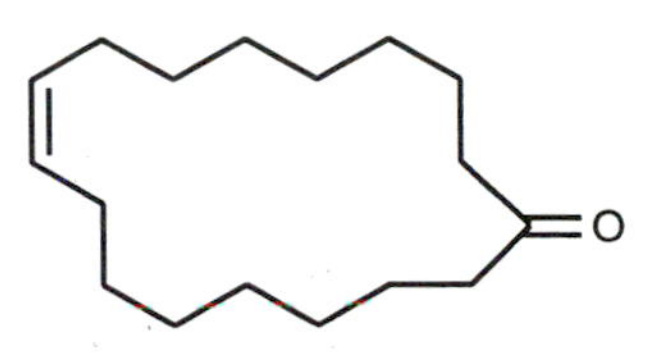

Ring size	5	6	11	17	34
% yield	80	80	0.2	8	2

However, today there are a wide range of alternative and more versatile methods for making rings. These will be classified according to the nature of the ring-forming processes involved.

3.4.2 Intramolecular nucleophilic substitution A typical nucleophilic substitution is a kinetically controlled process. The nucleophile can be a carbanion or a heteroatom thereby offering a convenient and general route into carbocyclic and saturated heterocyclic structures. The table below demonstrates how the reaction rate varies for cyclisation of a series of ω-bromoamines.

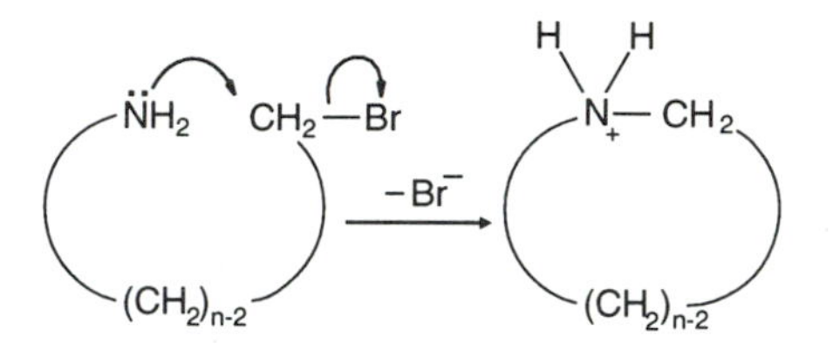

n	3	4	5	6	7	10	15
k_{rel}	0.12	0.002	100	1.7	0.028	10^{-8}	3×10^{-4}

Cyclic ethers can also be formed from ω-halo-alcohols in the presence of base. This provides a useful route to oxiranes (epoxides). These are both more readily formed and more reactive than their four-membered analogues (the trimethylene oxides) reflecting the fact that the activation energy for three-membered ring formation is lower than that for the four-membered ring though the latter are more thermodynamically stable making opening of the four-membered ring a significantly slower process.

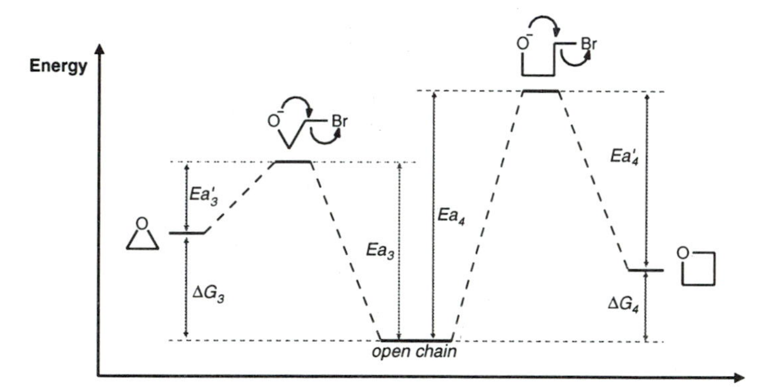

3-Membered rings are easier to form than their 4-ring counterparts but are more reactive. This reflects the fact that Ea_3 and Ea'_3 are less than Ea_4 and Ea'_4 respectively.

3.4.2.1 Bis-alkylation of Carbanions. Cyclopropane itself is readily formed by a variety of 1,3-displacement reactions (3-*exo-tet* processes) involving carbanions, such as those outlined below:

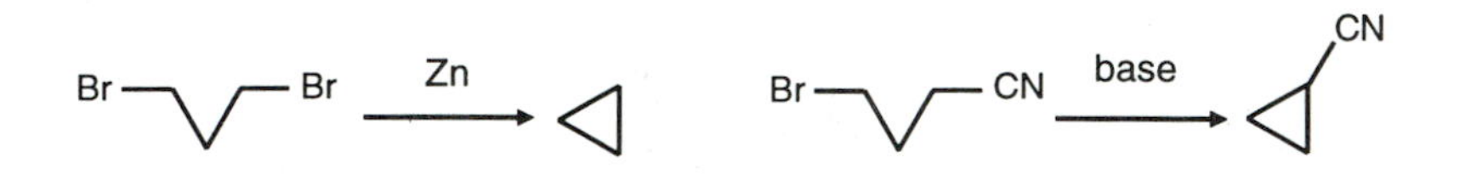

Reagents like diethyl malonate in which there is a methylene group substituted by two carbanion stabilising substituents (*i.e.* CH_2X_2 or CH_2XY, where X and Y are C(=O)R, COOR, CN, SO_2R, SR etc.) are particularly valuable for ring formation since both C-H bonds can be used to form new carbon-carbon bonds. For example, in the presence of two molar equivalents of sodium ethoxide, diethyl malonate can be alkylated twice. This process provides a convenient route to small and normal rings. Hydrolysis and decarboxylation of the resulting cyclic diester provides a convenient route to the mono-carboxylic acid.

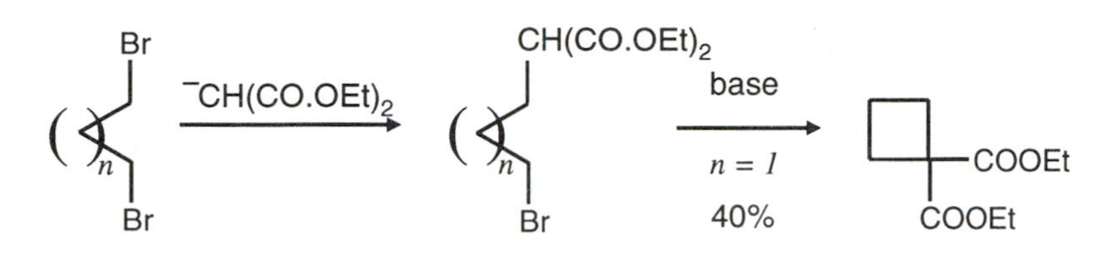

Three membered ring formation by *bis*-alkylation of diethyl malonate occurs *ca.* 10^5 times faster than formation of the cyclobutane analogue.

Allylic displacement also provides a useful route to cyclopropanes. For example, when diethylmalonate is reacted with 1,4-dibromo-*trans*-but-2-ene in the presence of 2 molar equivalents of base, initial alkylation and subsequent second deprotonation affords an intermediate anion which is geometrically constrained to undergo an S_N2' displacement leading to a vinylcyclopropane as product.

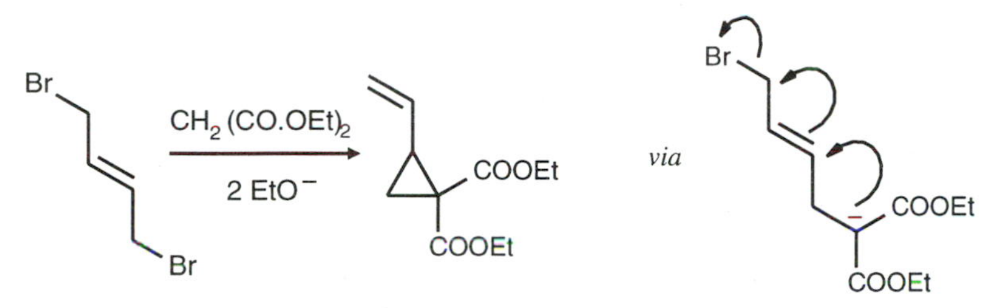

Acetylacetone (pentane-2,4-dione), ethyl cyanoacetate and malononitrile [$CH_2(CN)_2$] can be used in a similar manner.

Ethyl acetoacetate has the potential for alkylation at the central methylene group or on the ketonic methyl group. Both reactions can occur in a controlled fashion. The central methylene group is the most acidic centre in the molecule and in the presence of 1 molar equivalent of a base an anion is formed here. However, when ethyl acetoacetate is treated with excess (2 equivs.) of a very strong base such as sodamide ($NaNH_2$) or lithium di-isopropylamide [LDA, $LiN(CHMe_2)_2$] a second deprotonation can occur at the terminal methyl group to give a bis-enolate:

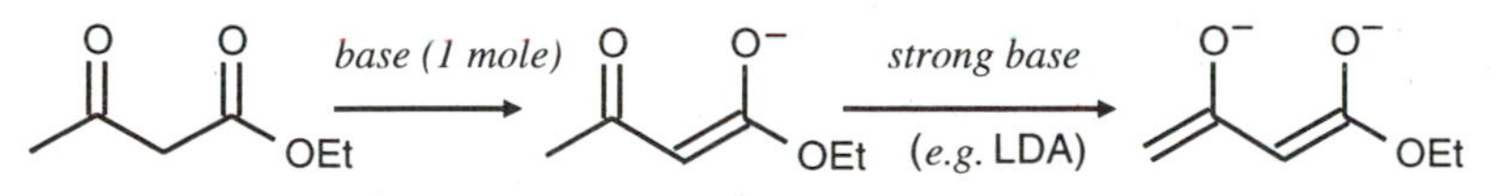

When this dianion is treated with 1,3-dibromopropane alkylation occurs initially at the more reactive terminal enolate and cyclisation onto the central carbon atom ensues giving a 6-membered ring.

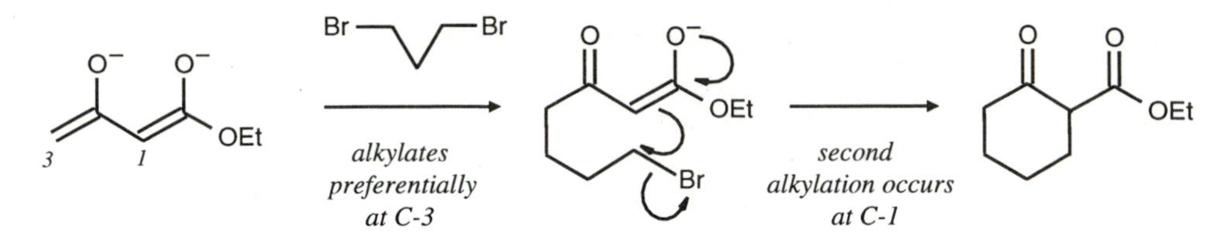

However, if excess of a weaker base such as sodium ethoxide is used instead, only the normal monoenolate is ever present and alkylation occurs twice on the methylene carbon to give 4-membered ring product.

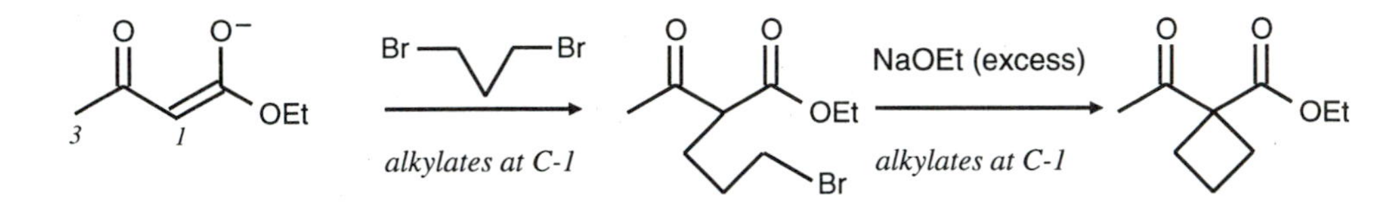

A more interesting reaction involves the use of 1,3-dithiane, a thioacetal of formaldehyde. The methylene hydrogens are much less acidic in this compound ($pK_a \sim 31$) and therefore deprotonation requires the use of a much stronger base such as BuLi. Reaction of the monoanion with α,ω-dihalides followed by reaction of the mono-alkylated intermediate with more base affords a spirocyclic thioketal which can be hydrolysed in the presence of an Hg^{2+} catalyst to give the cyclic ketone.

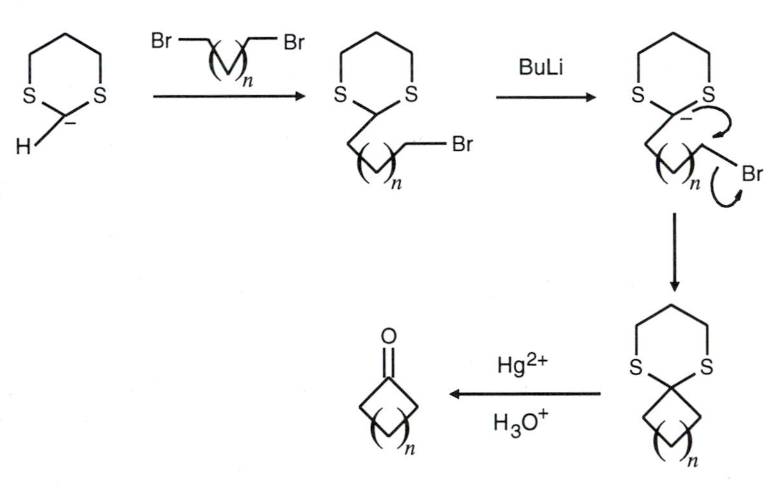

Best yields of cyclobutanone (*ca.* 50%) are obtained when 1-iodo-3-chloro-propane is used and the first substitution (of iodide) is carried out at low temperatures.

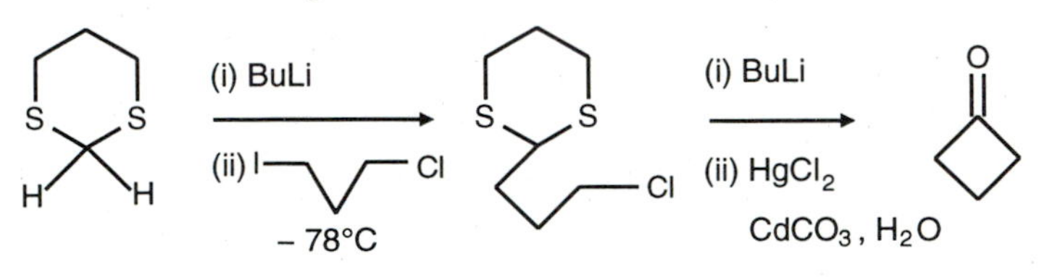

3.5 Nucleophilic Addition to unsaturated systems

3.5.1 The Thorpe-Ziegler reaction In the presence of a strong base such as $LiNEt_2$ (or LDA) α,ω-dinitriles cyclise to cyano-imines which can be subsequently hydrolysed to cyclic ketones. Under high dilution conditions this affords reasonable yields of large rings but fails for C_9 to C_{12} cycloalkanones.

Ring size	6	7	8	9	10	11	12	13	14
% yield	95	95	78	0	0	2	-	14	57

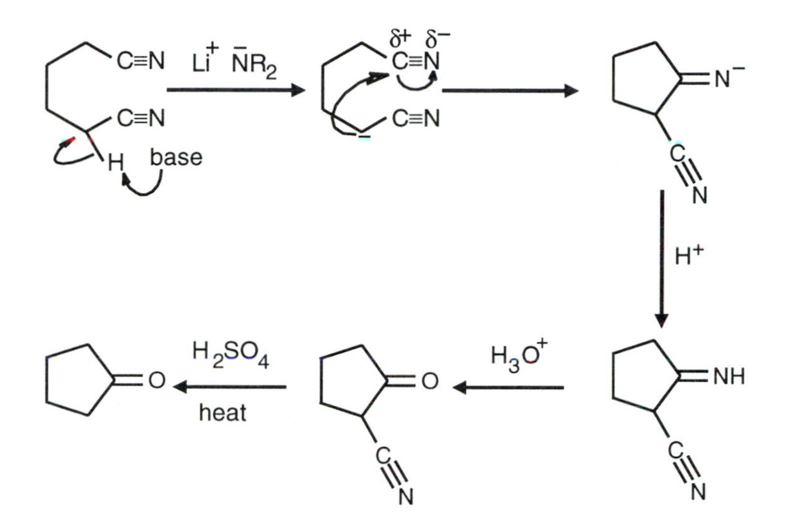

3.5.2 The Aldol condensation Nucleophilic addition of an enolate to another carbonyl group is a common condensation reaction for acyclic aldehydes and ketones. The reaction is reversible and is therefore thermodynamically controlled. Accordingly it can only be used to prepare normal rings.

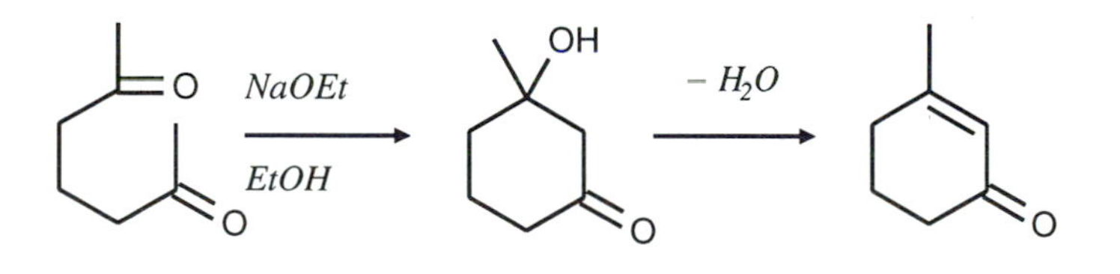

For many diketones there are often several alternative condensation pathways possible, but only those leading to 5- or 6-membered rings will occur. Thus 1,5- and 1,7-diketones both give 6-membered ring products and a 1,4-diketone forms a cyclopentenone with no evidence for formation of the alternative cyclopropyl ketone.

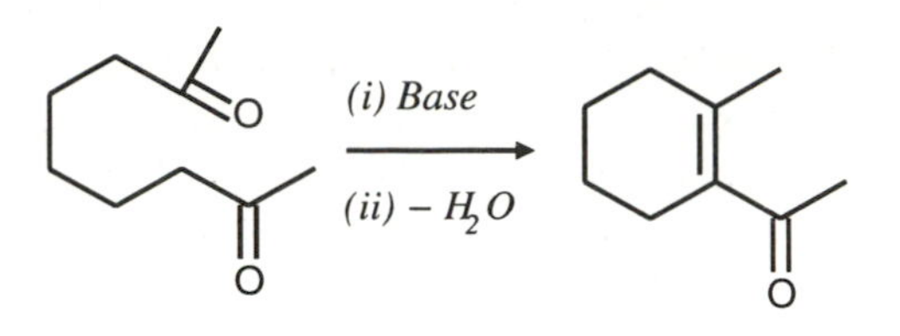

Unfavourable reaction pathways involving small- or medium-ring intermediates are avoided in these base-catalysed cyclisations.

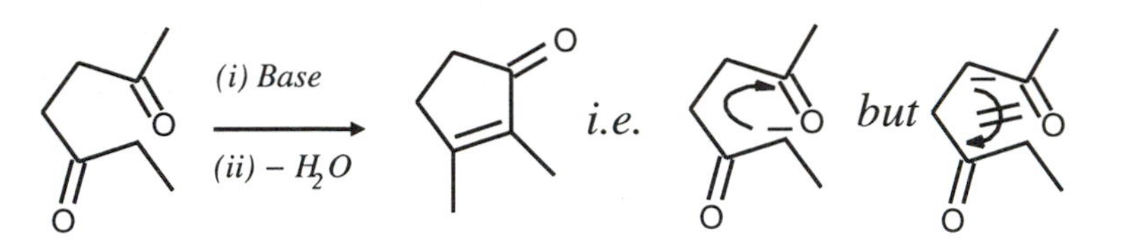

It is important to note that unsymmetrical diketones can give a mixture of cyclised products, but if only one of these can dehydrate, it is this that predominates.

3.5.3 Claisen condensation/Dieckmann cyclisation In the presence of base, esters undergo self-condensation to form β-ketoesters, the Claisen reaction. If the two ester groups are within one molecule, a ring can be formed (the Dieckmann reaction). This process is equilibrium controlled and is therefore generally particularly suitable for the preparation of 5-, 6- and 7-membered rings. It fails completely for small and medium rings (see Table below) but large rings can be isolated in moderate yields using high-dilution techniques.

Ring size	5	6	7	8	9	10	11	12	13	14
% yield	60 - 90	76	47	14	6	0	0	0	22	30

The reaction is driven by the formation of a resonance stabilised β-keto ester enolate intermediate. The preparation of ethyl cyclopentanone 2-carboxylate by reaction of diethyl adipate with sodium ethoxide in ethanol provides a good example of the process.

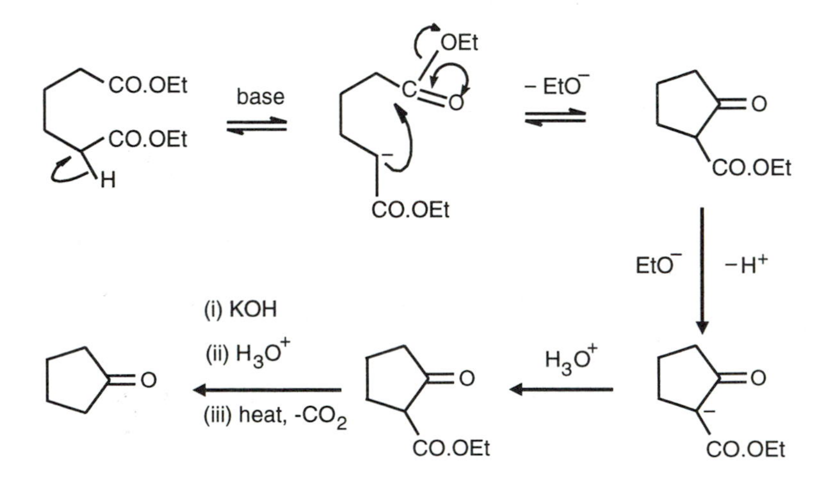

Yields can be significantly improved through the use of a strong base such as NaH or Ph_3C^-, which makes the initial deprotonation irreversible. For example, reaction of diethyl suberate (EtOCO-$(CH_2)_6$-COOEt) with sodium ethoxide gives no cycloheptanone derivative as product whereas when stronger bases like KO*t*Bu or NaH are used the seven-membered ring is obtained in 47-58% yield.

3.5.4 Michael addition - the Robinson ring annelation Carbanions can also add to activated alkenes particularly α,β-unsaturated carbonyl derivatives; this provides a useful approach to ring formation. For example, reaction of a dienone with diethyl malonate in the presence of base results in two sequential Michael additions giving a 4,4'-disubstituted cyclohexanone as product.

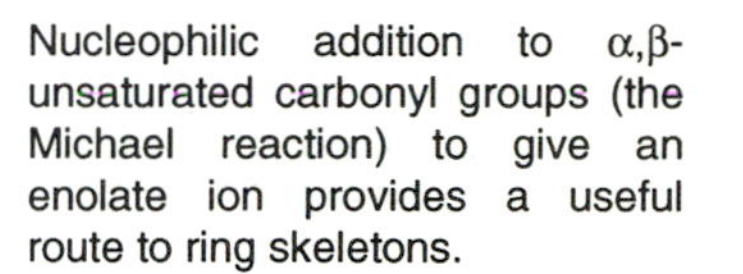
Nucleophilic addition to α,β-unsaturated carbonyl groups (the Michael reaction) to give an enolate ion provides a useful route to ring skeletons.

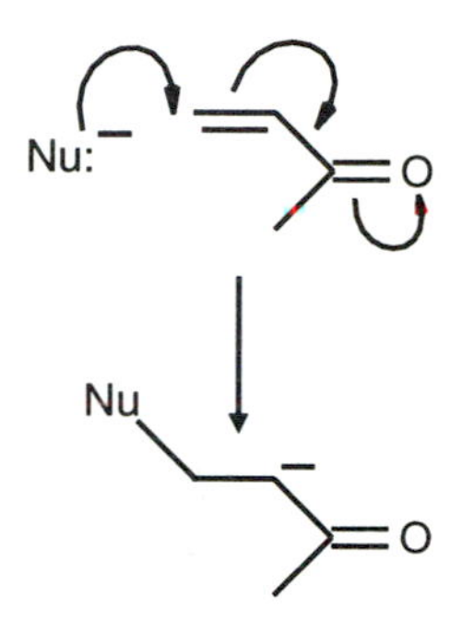

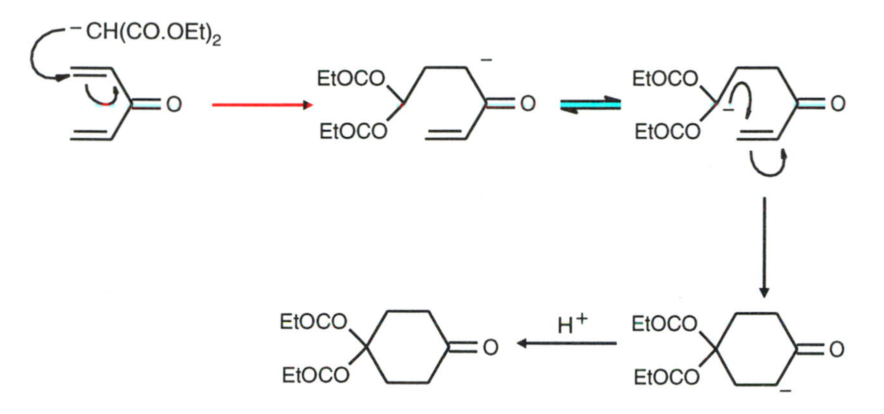

When methyl chloroacetate is reacted with methyl acrylate in the presence of sodium methoxide, Michael addition leads to an enolate intermediate which can "kick back" and irreversibly displace chloride resulting in formation of a cyclopropane.

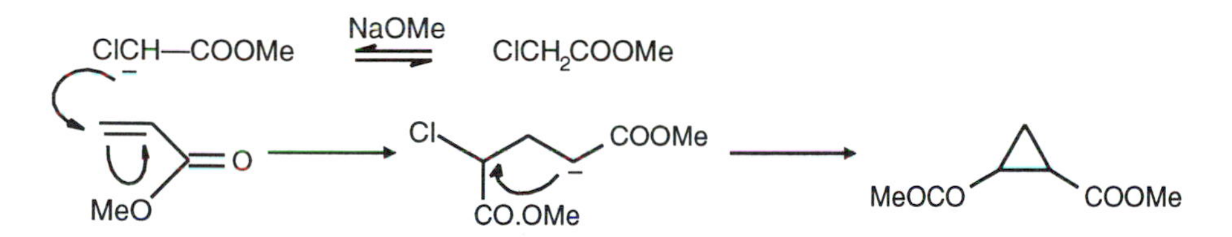

In a similar manner, Michael addition using a sulfone-stabilised carbanion affords a simple synthesis of chrysanthemic acid:

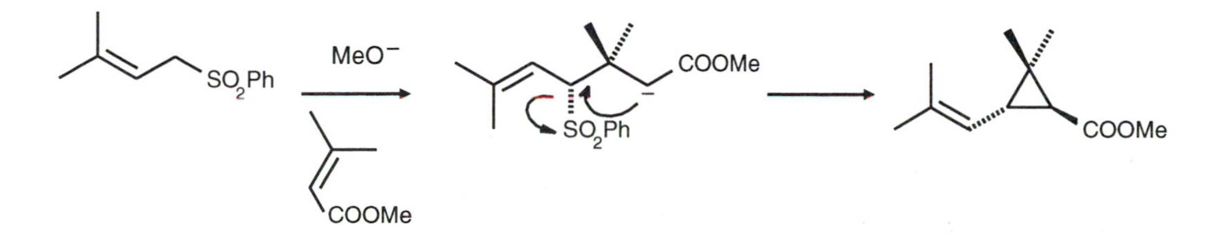

Perhaps the most important example of the use of the Michael addition is in the synthesis of fused cyclohexenones, a process known as the Robinson ring annelation. This has provided a valuable route to steroidal derivatives. It involves the reaction of a 2-substituted cyclohexanone with methyl vinyl ketone in the presence of base. Michael addition is followed by base-catalysed condensation giving a cyclohexenone product.

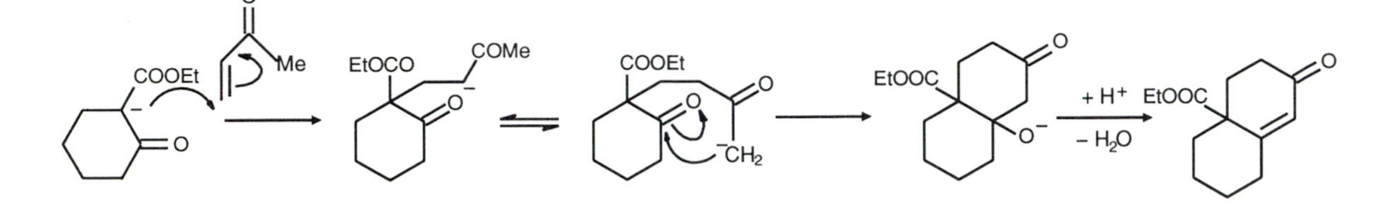

However, the basic conditions required for the Robinson reaction tend to induce polymerisation of the vinyl ketone resulting in poor yields. In addition the Michael adduct and the starting ketone have a similar pK_a leading to competitive reactions for the electrophile. (Significantly improved yields are obtained through the use of α–silylated vinyl ketones which react with lithium enolates to give a Michael adduct. This can then be cyclised by treatment with sodium methoxide.)

When the Robinson reaction is applied to larger ring cyclic ketones, bridged bicyclic structures frequently result. This alternative pathway predominates for cyclo-octanones and larger ketones, the carboalkoxy group being lost during the course of the reaction.

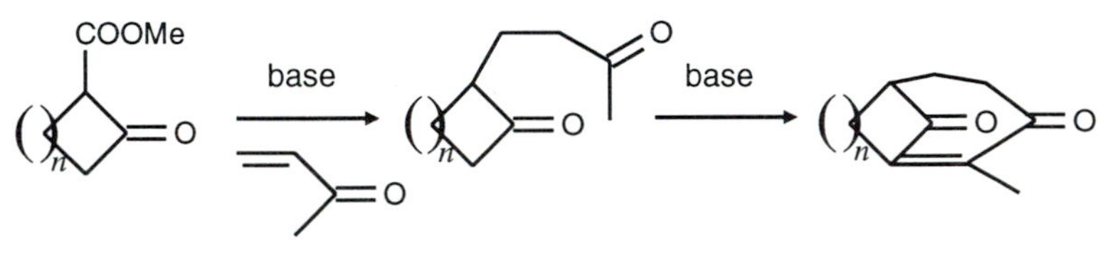

3.5.5 Other cyclisation reactions involving carbanions

3.5.5.1 Sulfoxonium ylids. Whilst sulfoxonium ylids react with ketones to form epoxides their behaviour towards α,β-unsaturated carbonyl compounds is somewhat different, Michael addition being the preferred reaction pathway. The intermediate enolate then fragments by displacement of a sulfoxide giving a cyclopropane derivative. The reaction occurs even in highly hindered systems such as that shown below:

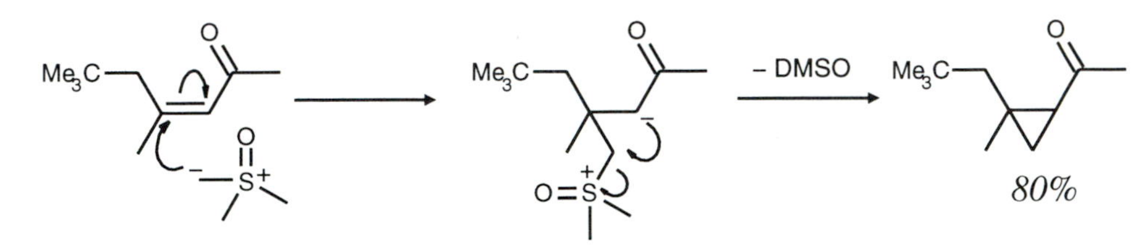

Another valuable sulfur ylid is that derived from cyclopropyldiphenylsulfonium tetrafluoroborate. This reacts with aldehydes and ketones to form an oxaspiropentane, which can undergo acid-catalysed rearrangement (or more cleanly with lithium tetrafluoroborate) to give a cyclobutanone. Reaction of this ylid with cyclohexanone affords the corresponding spiro cyclobutanone:

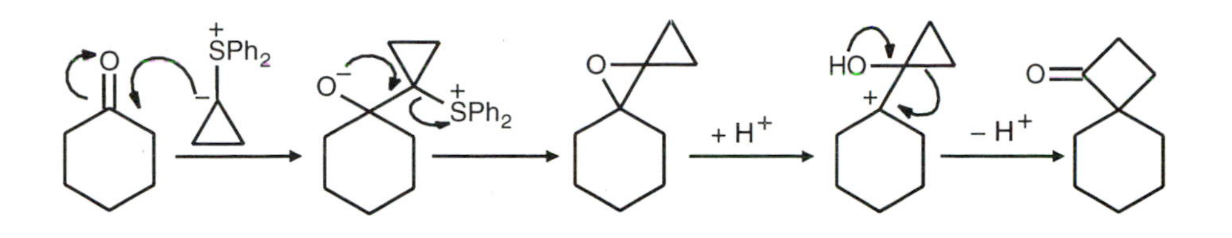

3.5.5.2 Phosphorus ylids - The Wittig reaction. This provides an important route for converting aldehydes and ketones to alkenes. If the carbonyl group and ylid are in the same molecule an intramolecular reaction can lead to formation of a cycloalkene.

Examples of this process involving both phosphoranes and phosphonate stabilised carbanions (the Wadsworth Emmons reaction) are known. The latter has been used to make large macrocyclic lactones related to naturally occurring antibiotics:

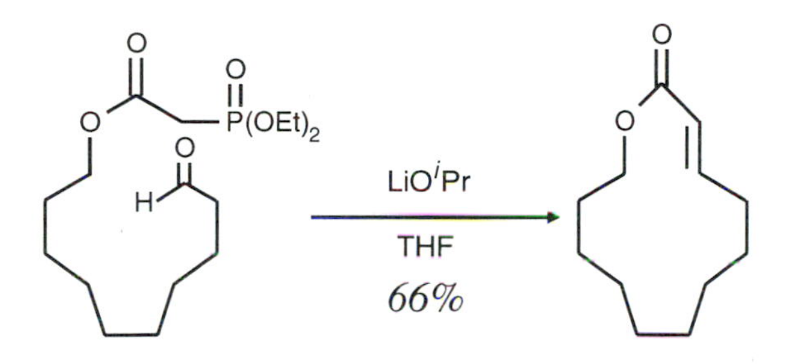

Normally reaction between ylids and esters is unfavourable, but intramolecular substitution involving reaction between an ylid and an ester (*c.f.* a Dieckmann reaction) gives cycloheptanone in good yield:

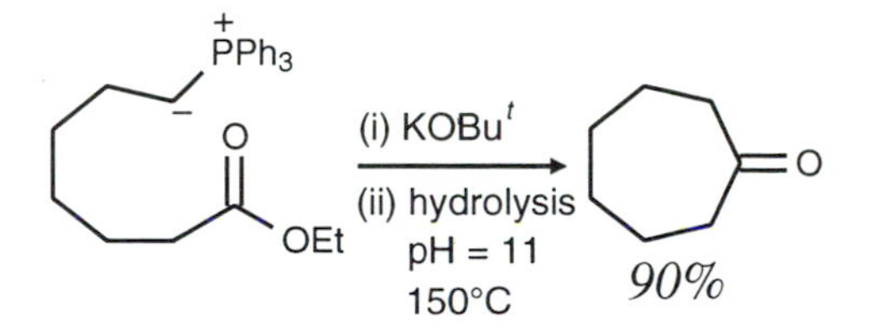

3.6 Electrophilic addition to unsaturated systems

We have already noted that alkenes are susceptible to electrophilic attack. When the electrophile is a carbocation, a carbocyclic ring can result, an important process in natural systems, particularly the terpenes.

For example, under acidic conditions linalool is converted to α–terpineol.

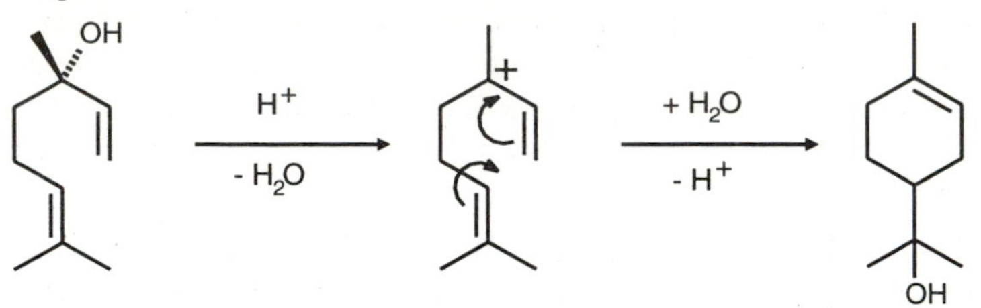

Such reaction pathways normally lead to cyclohexane derivatives, and in multiple cyclisations this process occurs with impressive control of stereochemistry arising from the involvement of a low-energy chair conformation for the cyclisation transition state.

Squalene cyclisation:

Sterols are formed in biological systems by the acid-catalysed multiple cyclisation of squalene 2,3-oxide. This remarkable process generates 7 new chiral centres but only one of the 128 possible isomers is formed:

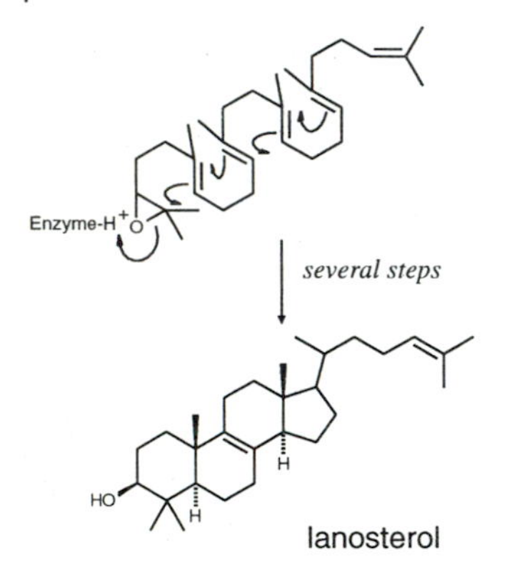

Such stereoelectronic control is evident in the enzyme-induced multiple cyclisation of the polyene, squalene, which results in the formation of a steroidal skeleton with remarkable stereospecificity.

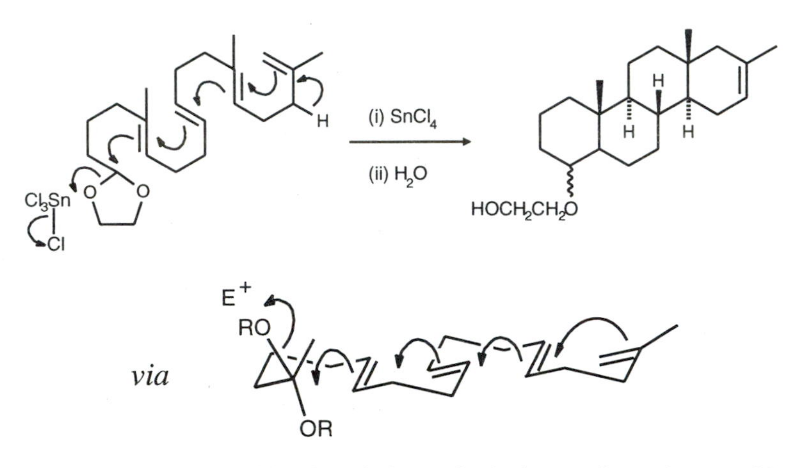

π-Electron participation during solvolysis reactions also provides a useful route to bi- and polycyclic structures. This is well illustrated by the formation of 2-norbornyl acetate from acetolysis of (3-cyclopentenyl)ethyl sulfonate esters ($X = ArSO_3$ –):

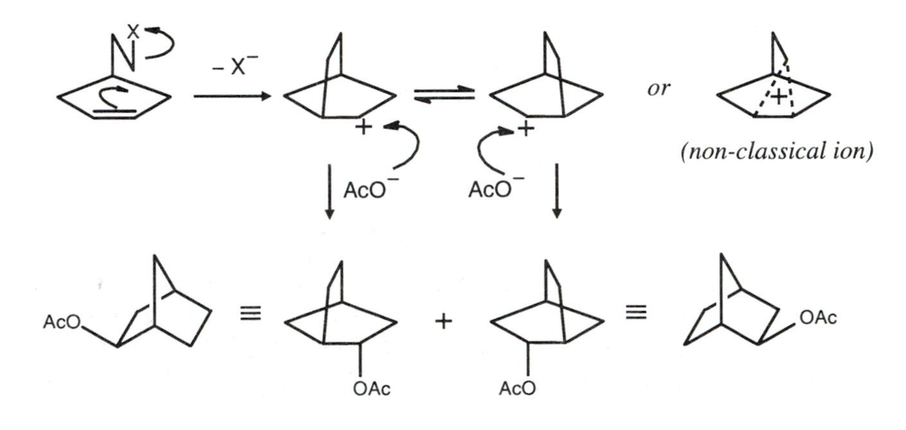

3.7 Pericyclic and related processes

Pericyclic reactions of various types provide a valuable method of making ring structures of between 3 and 6 atoms. During such processes two skeletal ring bonds are formed in one transition state and this leads to potentially valuable control of stereochemistry of a number of centres about the ring.

Addition of a singlet-state carbene to an alkene is a concerted reaction which may be considered as a cycloaddition resulting in stereospecific cyclopropane formation:

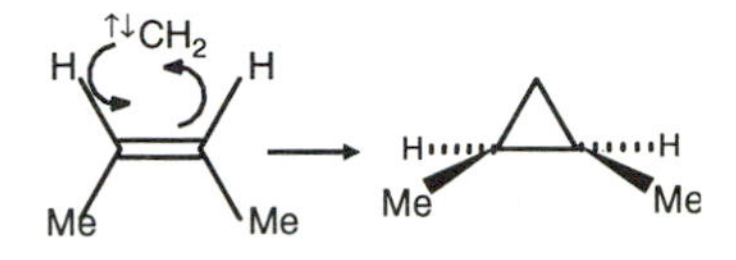

whereas triplet carbenes behave as diradicals and add in a stepwise manner leading to stereorandomisation:

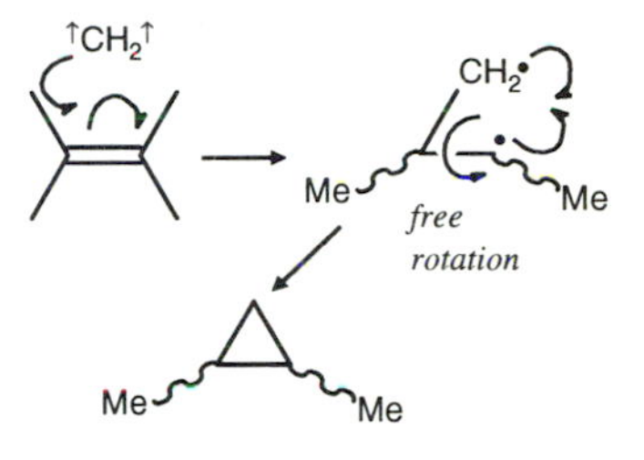

3.7.1 Addition of carbenes and carbenoids to alkenes Three-membered rings are readily formed by the addition of carbenes and carbenoids to alkenes. Relatively stable singlet carbenes such as dichlorocarbene :CCl_2 (which is readily generated by reaction of chloroform with base or the thermal decomposition of sodium trichloroactetate) add across alkenes in a concerted fashion to give cyclopropanes in which the alkene stereochemistry is retained. Bicyclic structures are also readily formed in this way. Though dichlorocarbene gives a dihalocyclopropane, the halogen substituents are readily replaced by alkyl groups using organocuprates (*e.g.* $LiCuMe_2$ which converts >CCl_2 into >CMe_2) or may be reduced to >CH_2 (by Bu_3SnH in the presence of a radical initiator).

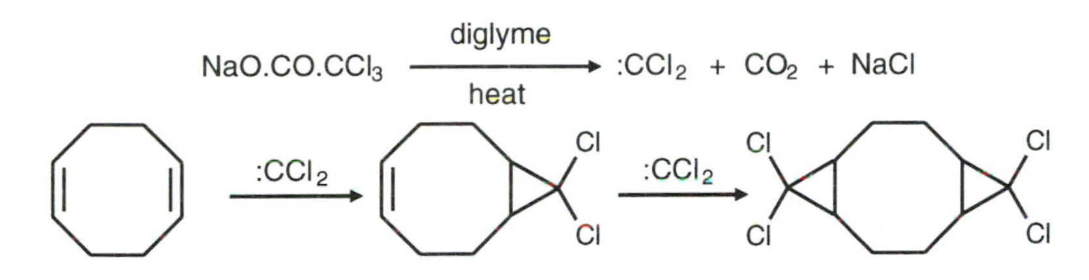

Dichlorocarbene will also add across an alkyne to produce a dichlorocyclopropene. Such compounds are readily hydrolysed to cyclopropenones, a process which is facilitated by the intermediacy of an aromatic cyclopropenium cation.

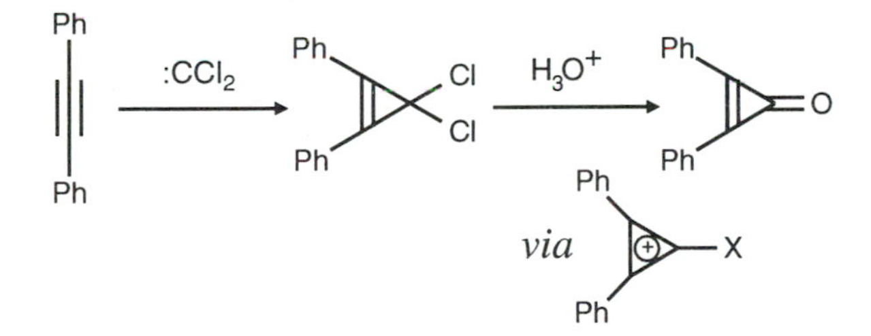

The addition of methylene, :CH_2 requires a different approach because the high reactivity of the free carbene leads to many side reactions particularly insertion into σ-bonds. A particularly useful and controlled method for adding methylene across alkenes to form cyclopropanes in good yields involves the Simmons-Smith procedure. Here di-iodomethane is treated with a zinc/copper couple (prepared from Zn dust and CuCl) in the presence of an alkene. The likely reactive intermediate is an organozinc complex *e.g.* $(ICH_2)_2Zn.ZnI_2$ and addition

The Simmons-Smith reaction:

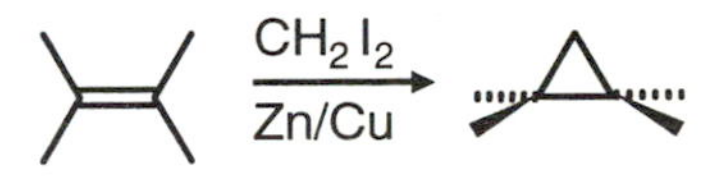

is stereospecifically *syn* supporting a concerted pathway. This is a versatile reaction which can be carried out in the presence of a wide variety of functional groups. For example, the methyl ester of dihydrosterculic acid is obtained in 51% yield from methyl oleate:

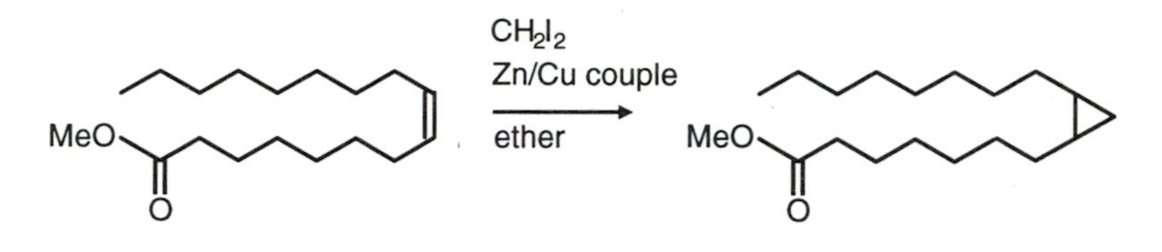

Diazoalkanes readily lose nitrogen to form carbenes on photolysis or by Cu-catalysed decomposition, the two processes giving different product distributions. For example the Cu-catalysed decomposition of α-diazoketones and esters provides a useful route to cyclopropane derivatives:

Diazoalkanes form carbenes (or carbenoid intermediates) upon photolysis or by copper-catalysed decomposition.

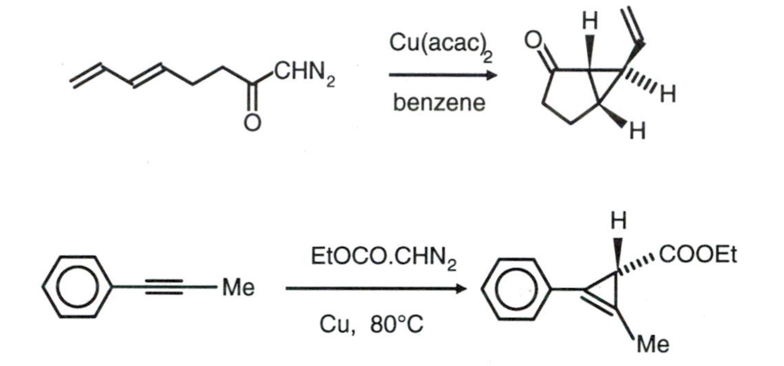

3.7.2 Cycloaddition reactions

Diazoalkanes undergo 1,3-dipolar cycloaddition reactions with alkenes:

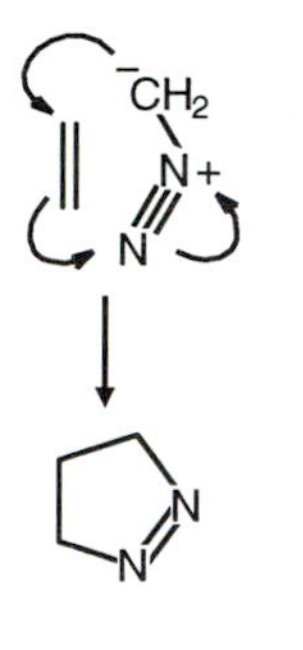

3.7.2.1. 1,3-Dipolar cycloadditions. Diazoalkanes are 1,3-dipolar in character and will undergo cycloadditions with alkenes to give pyrazolines. These extrude nitrogen on thermolysis or photolysis to give a 1,3-diradical which then cyclises to the corresponding cyclopropane. For example, reaction of ethyl diazoacetate with *trans*-cyclo-octene, followed by thermal loss of N_2 provides a useful route route to *trans*-bicyclo[6.1.0]nonyl esters. Whilst the thermal decomposition frequently leads to stereoisomeric mixtures of cyclopropane products, the *trans*-bicyclo[6.1.0]nonane is favoured, being the (slightly) more stable stereoisomer.

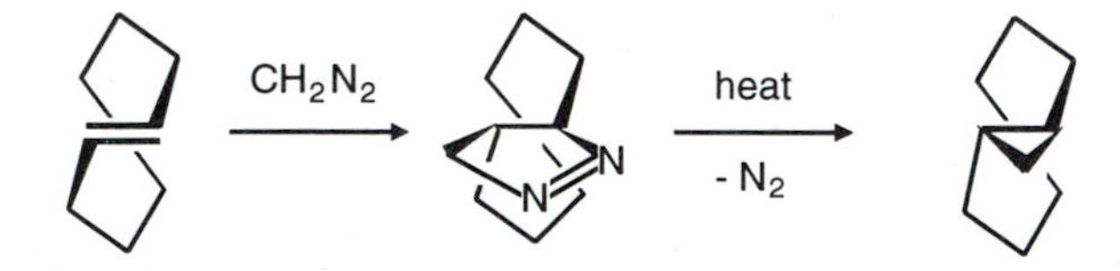

3.7.2.2 [2+2]-Cycloadditions. Whilst the cycloaddition of one alkene with another is a "thermally-forbidden" process, the reaction does occur in a stereospecific manner in the presence of light (*trans*-but-2-ene gives only two of the possible stereoisomeric tetramethylcyclobutanes, there being two possible [2+2]-cycloaddition transition-state geometries).

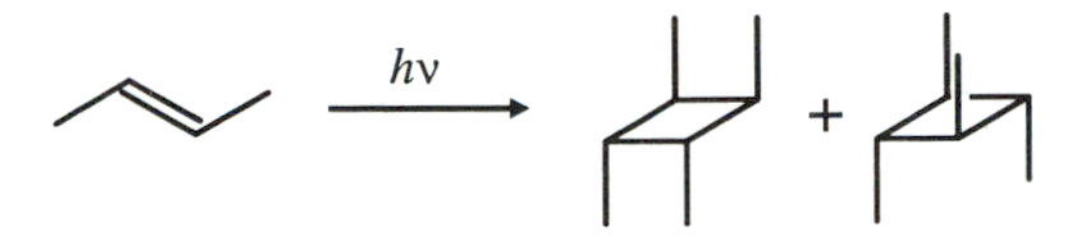

[2+2]-cycloadditions are thermally forbidden but photochemically allowed as is evident from consideration of the HOMO-LUMO interaction between two ethene molecules under both sets of conditions:

Thermal reaction: (both molecules in ground state; HOMO = π LUMO = π*) - "forbidden"

Photochemical reaction: (one molecule in ground state, LUMO = π*; one molecule in excited state, HOMO = π*) - "allowed"

Such reactions can occur intramolecularly to give highly strained products, *e.g.* the formation of quadricyclane from photoisomerisation of norbornadiene:

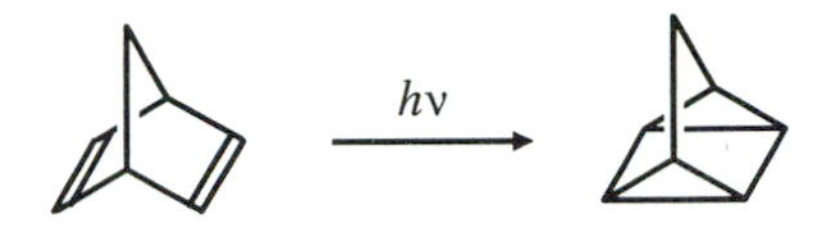

Perhaps more generally synthetically useful transformations involve the photo-initiated addition of enones across alkenes and the thermal [2+2]-cycloaddition of ketenes to alkenes.

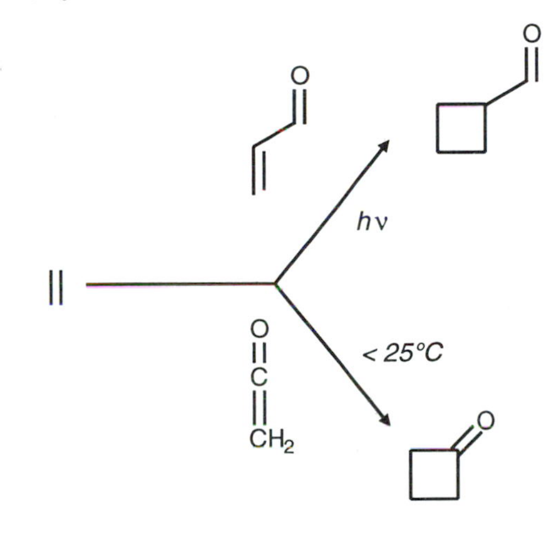

The cycloaddition of enones to alkenes is not a concerted process but rather operates through a stepwise pathway in which the enone photoexcited state behaves as a diradical. Such processes can occur intramolecularly leading to the formation of complex polycyclic structures such as that of carvonecamphor from photoisomerisation of carvone:

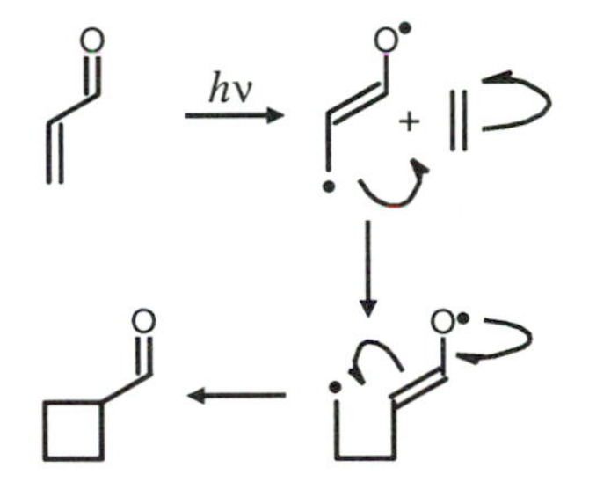

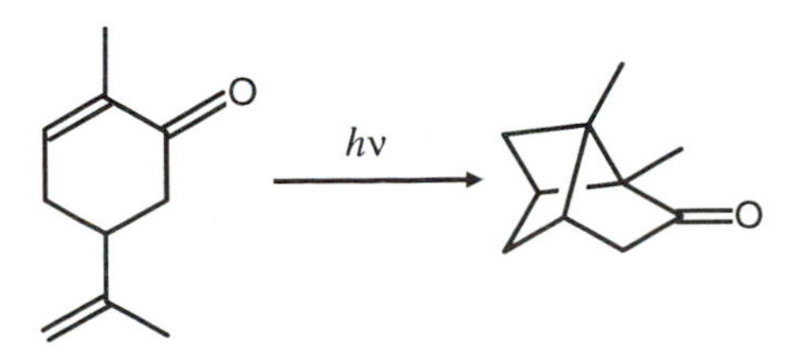

The {2+2}-cycloaddition of ketenes with alkenes may be considered as proceeding by a thermally allowed suprafacial-antarafacial addition pathway:

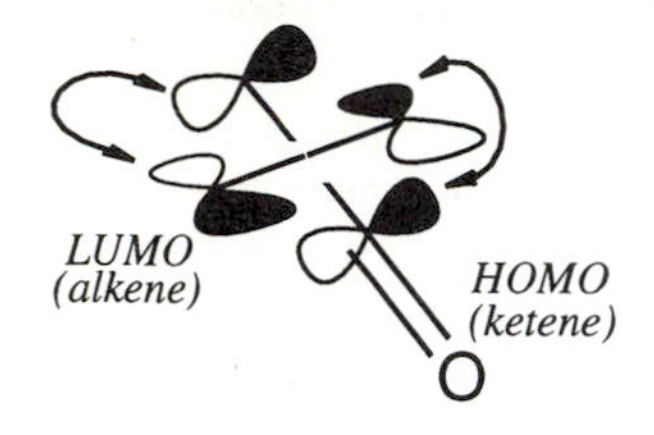

Such a process is severely sterically hindered in simple alkenes.

Ketenes react efficiently with alkenes and conjugated dienes by a {2+2}-cycloaddition pathway at room temperature to give cyclobutanones. The preference for a [2+2] cycloaddition by ketenes is well-illustrated by their reaction with dienes, where a Diels Alder reaction could compete but four-membered ring formation is the exclusive reaction pathway. For example, dichloroketene which is readily generated by the Zn-induced dehalogenation of trichloroacetyl bromide, adds to cyclopentadiene to give a bicyclo[3.2.0]heptenone.

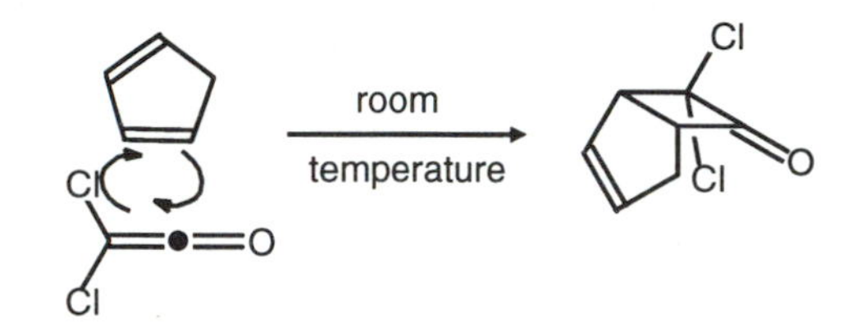

However, the [4+2]-cycloaddition (a Diels-Alder process) of ketenes with dienes would provide a valuable synthetic route to cyclohexenes and bridged rings and this has been achieved through the use of "masked" ketenes, *i.e.* compounds which are the synthetic equivalents of ketenes (see below).

3.7.2.3 The Diels Alder reaction. A particularly useful example of a ketene equivalent is 2-acetoxy-acrylonitrile, reaction of which with, for example, cyclohexadienes provides a useful route to bicyclo[2.2.2]-octanone derivatives:

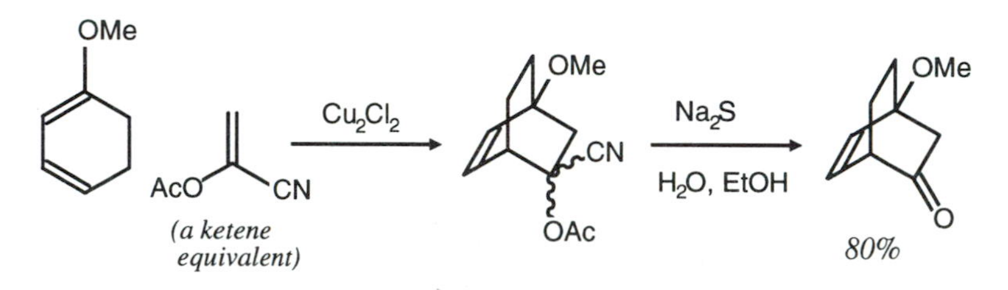

In general the [4+2]-cycloaddition provides a valuable synthetic route to six-membered rings particularly because of the potential control of stereochemistry it offers. In the simplest case the reaction is favoured by the presence of electron-withdrawing groups on the alkene (*i.e.* the dienophile) and proceeds in a suprafacial/suprafacial manner leading to "cis" addition of the alkene across the diene framework. This results in the preservation of stereochemical information from both the alkene and the diene as shown in the following example:

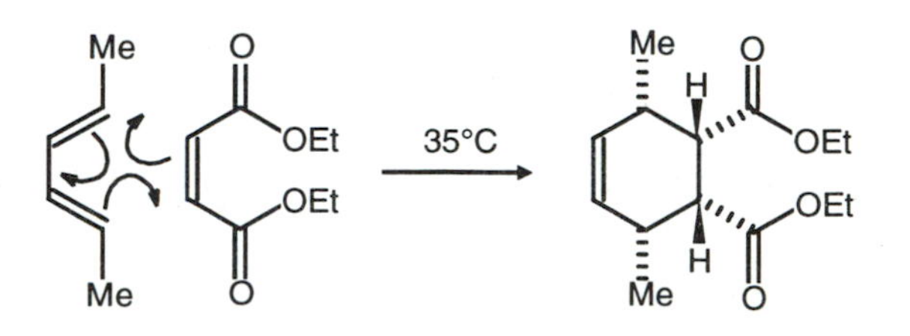

One key feature of the Diels Alder reaction is highlighted by reactions leading to bicyclic structures such as the norbornenyl (bicyclo[2.2.1]heptenyl) adduct formed between maleic anhydride and cyclopentadiene. At low temperatures the *endo* product is formed (under kinetic control) but the cycloaddition is reversible and if the *endo* product is heated it isomerises to the thermodynamically more favourable *exo*-isomer. The kinetic preference for the *endo* product, which is a general observation in such reactions, can be understood as arising from additional, favourable orbital interactions between the HOMO of the diene and the π–components of the LUMO of the dienophile on the carbonyl-group carbon atoms.

The terms *exo* and *endo* relate to the location of substituents about a bicyclic skeleton *e.g.* at C_2, C_3, C_5 or C_6 of a norbornane skeleton, *exo* being on the same face as the shortest bridge *(i.e.* C_7).

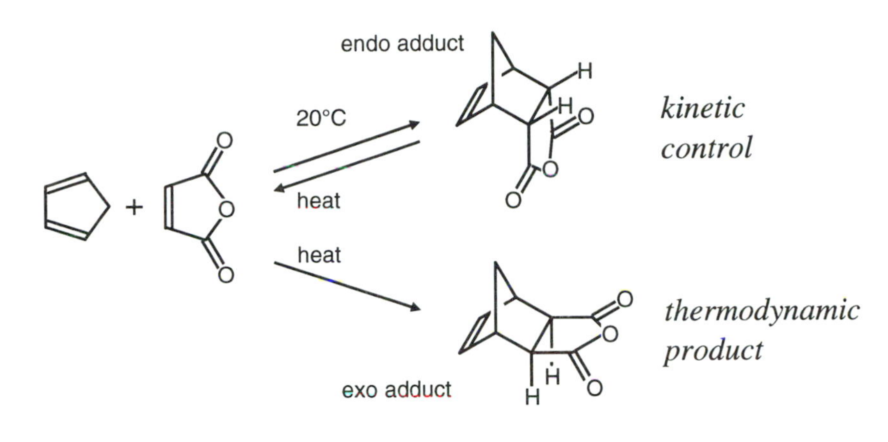

The relative stereochemistries of substituents of both diene and dienophile are retained in the Diels Alder adduct, which considerably enhances the synthetic value of this process. In particular, intramolecular Diels Alder reactions provide an important method for the synthesis of polycyclic systems such as those present in terpenes, steroids and alkaloids. Such reactions can be used to form several chiral centres having predetermined configurations in one step:

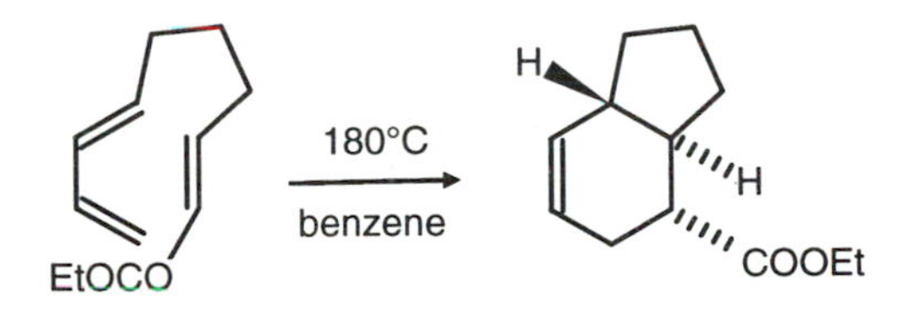

3.7.3 Electrocyclic ring closures Polyenes undergo thermal (Δ) or photochemically (hν) induced ring closure to give cyclic polyene derivatives in a stereochemically controlled manner. Ring closure can be conrotatory (CON) or disrotatory (DIS) depending on the number of π-electrons involved in the reaction transition state and the conditions (Δ or hν) employed.

The cyclisation of butadiene is a simple electrocyclic process:

Shown below are the HOMO and LUMO of buta-1,3-diene. The arrows indicate the bond rotation required to achieve a bonding interaction between the terminal orbitals of the polyene chain.

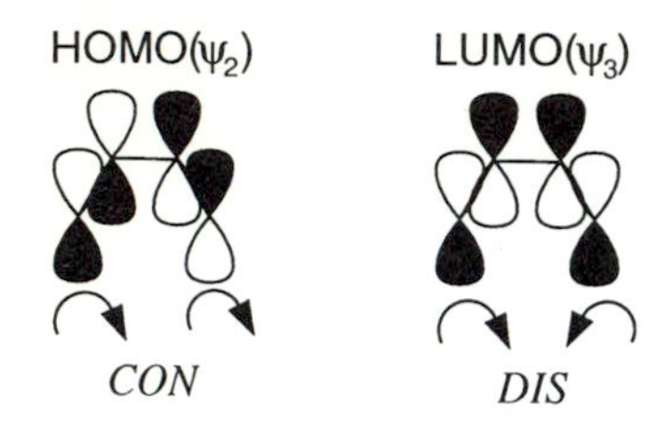

Buta-1,3-diene can in principle undergo either a thermally (conrotatory) or a photochemically (disrotatory) induced electrocyclic ring closure to form a cyclobutene. Whilst both processes are reversible, in practice the relative instability of the cyclobutene ring favours the thermal ring opening process, whereas the λ_{max} values for the alkene (*ca.* 190 nm) and the diene (*ca.* 220 nm) are such that under normal irradiation conditions only the diene absorbs light, leading to a photostationary state which favours cyclisation.

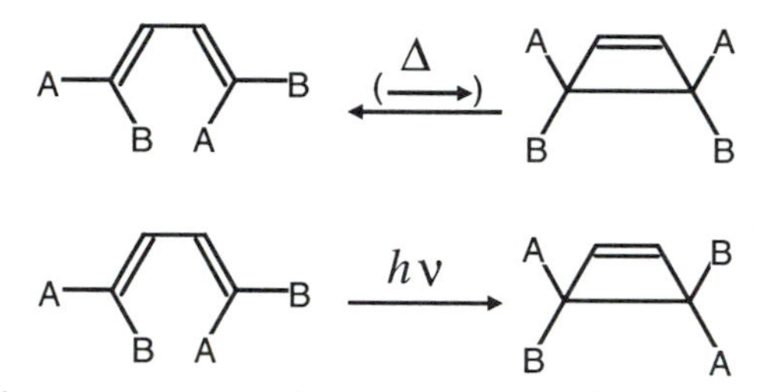

One important synthetic consequence of the stereospecificity of the ring-opening process arises when the cyclobutene is *cis*-fused to another ring since conrotatory ring opening generates a *cis,trans*-diene:

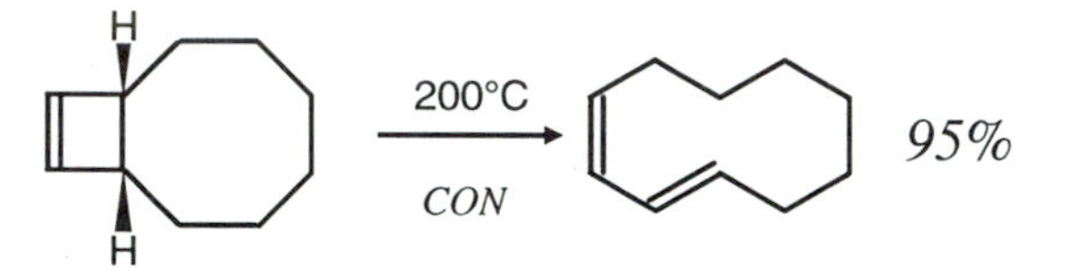

whereas thermal opening of the *trans*-fused isomer produces the isomeric *cis,cis*-diene.

3.7.4 Multiple cyclisations using pericyclic reactions As has already been noted, a particularly elegant use of pericyclic reactions involves the formation of several carbocyclic rings in one reaction step, a process which can be carried out with considerable stereochemical control. Combinations of electrocyclic ring opening of a benzocyclobutene to an *o*-xylylene followed by Diels Alder cyclisation have been successfully exploited in the synthesis of steroidal derivatives:

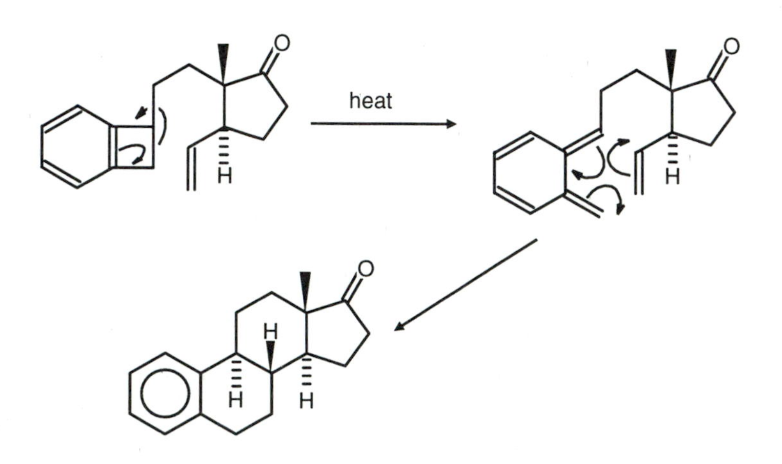

3.8 Electron-transfer processes

3.8.1 The acyloin reaction This is one of the most important synthetic routes to medium rings but can also be used to make other ring sizes. When the carbonyl group of an ester is reduced by sodium metal, a radical anion is formed which couples with another to give a dimer. If the two esters involved are within the same molecule a ring can be formed.

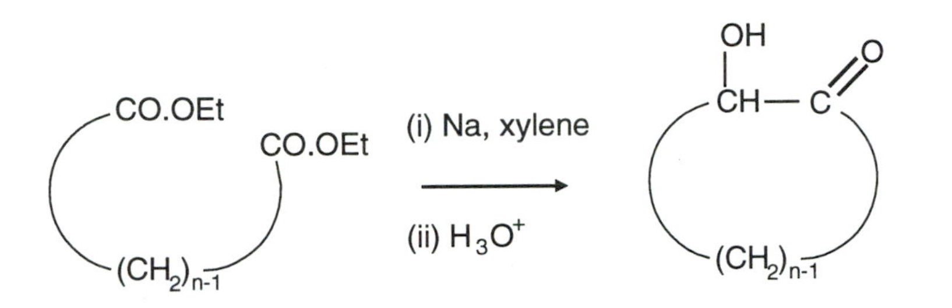

The acyloin reaction occurs on the surface of sodium metal. This favours cyclisation over polymerisation and helps the preparation of unfavourable ring sizes.

This reaction is carried out using sodium metal rapidly stirred in a boiling inert solvent such as xylene. Under such conditions the sodium is dispersed as fine molten droplets and therefore has a very high surface area. The reaction takes place on the metal surface which acts as a template bringing the two ester groups together. This reaction provides the best route to carbocyclic systems having 10 or more atoms.

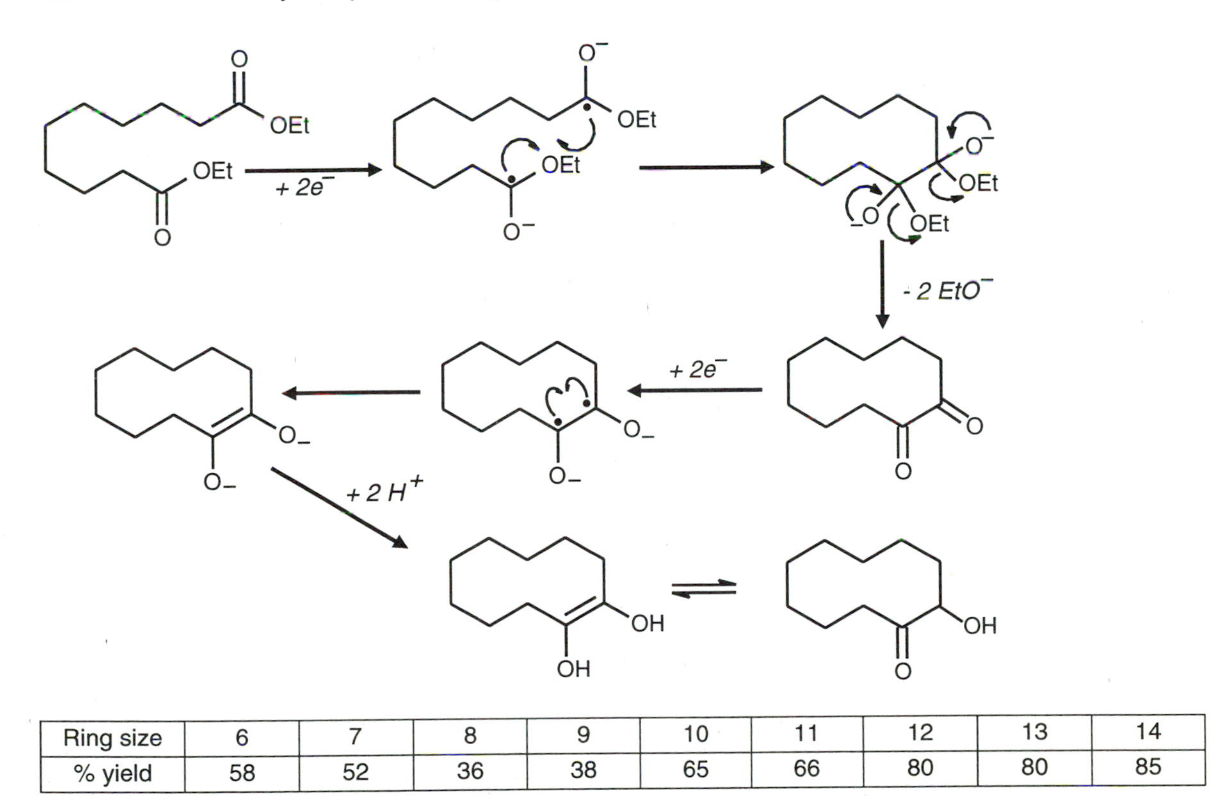

Ring size	6	7	8	9	10	11	12	13	14
% yield	58	52	36	38	65	66	80	80	85

One problem which can occur in this reaction arises from the presence of ethoxide ion. This can act as a base causing competing side reactions such as Claisen condensation which is a particular disadvantage when trying to use this procedure for the synthesis of small rings. This is effectively overcome by carrying the reaction out in the presence of trimethylsilyl chloride which scavenges the ethoxide ion and at the same time traps and protects the enediolate product. For example, in the presence of trimethylsilyl chloride, good yields of four-membered ring acyloin are obtained and those of other rings sizes are improved. When diethyl succinate is reacted with Na metal in xylene, Dieckmann cyclisation gives the dimer, 2,5-*bis*-(carboethoxy)-cyclohexane-1,4-dione, but when Me_3SiCl is present, the acyloin product 2-hydroxycyclobutanone is obtained in 70% yield. Under the same conditions diethyl suberate [EtO.CO–$(CH_2)_6$–CO.OEt] gives 2-hydroxycyclo-octanone (in 75% yield)

The Claisen condensation often competes with acyloin coupling.

However, when the acyloin reaction is carried out in the presence of trimethylsilyl chloride, side reactions are suppressed leading to improved product yields and the formation of small-ring acyloins which are otherwise inaccessible.

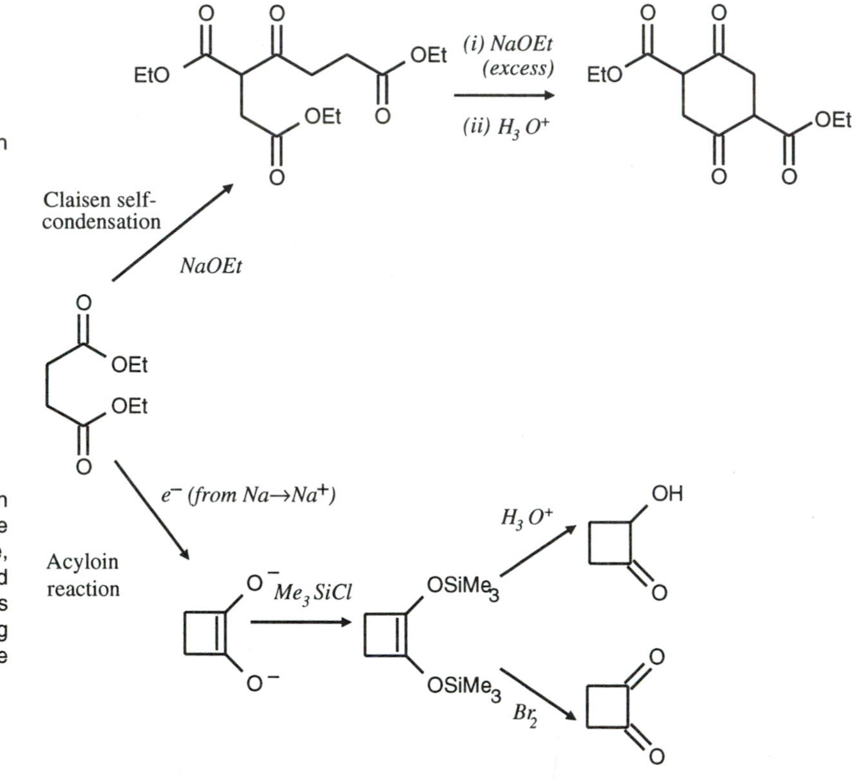

3.8.2 The pinacol reaction Dissolving metal reduction of ketones using magnesium amalgam, in the absence of a proton source, leads to coupling of carbonyl group radical anion intermediates affording dimeric 1,2-diol products (pinacols). Thus the reduction of cyclopentanone affords a pinacol which, on exposure to acid undergoes a rearrangement to a pinacolone product, in this case a spiro-cyclic ketone.

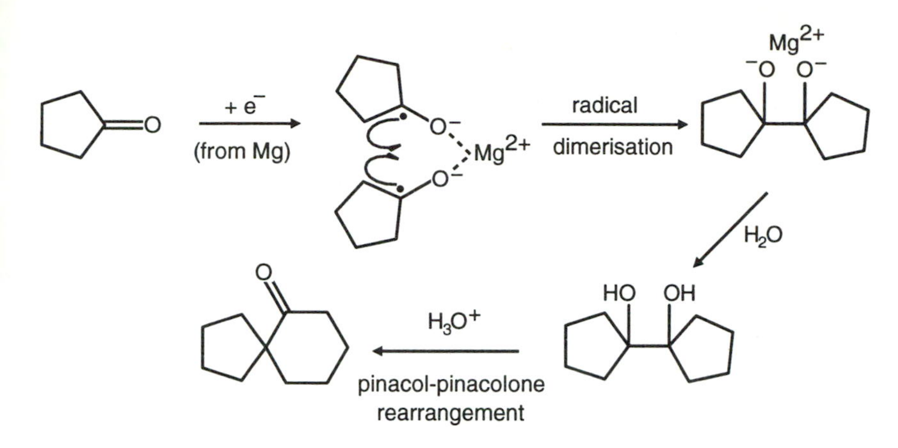

Recently the synthetic potential of the pinacol reaction has been considerably enhanced through the use of reagents derived from low-valent transition metal species, *e.g.* Ti(II), generated by reaction of $TiCl_4$ with Mg/Hg. Use of this reagent causes intramolecular pinacol reactions to proceed in excellent yield, even allowing the formation of cyclobutane-1,2-diols:

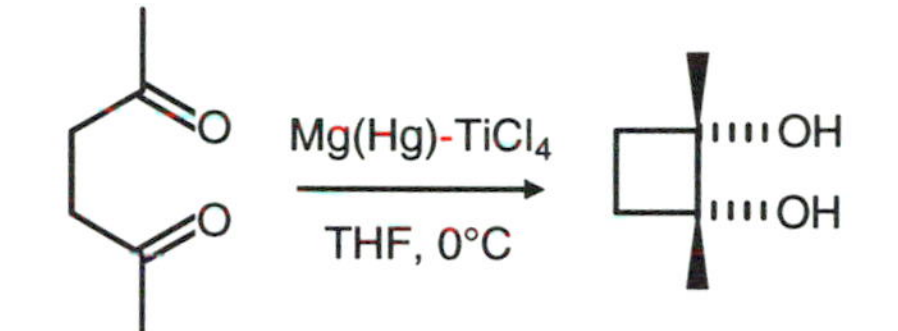

Reductive coupling of aldehydes and ketones using activated titanium species formed by reduction of $TiCl_3$ (the McMurray reaction) has been used to form cyclic alkenes having between 4 and 16 carbon atoms as illustrated by the synthesis of the cyclic diterpene flexibilene shown below:

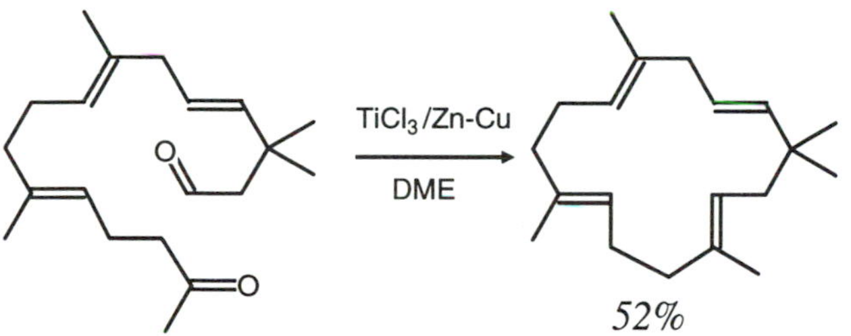

3.8.3 The Birch reduction Dissolving metal reduction of aromatic rings can be brought about using alkali metal-liquid ammonia solutions. This reaction proceeds under kinetic control to give the non-conjugated cyclohexa-1,4-diene (only conjugated dienes are further reduced under such conditions). Whilst benzoic acids are readily reduced by simple alkali metal/liquid ammonia solutions, benzene itself and electron-rich aromatic derivatives require the presence of ethanol to drive the initial, and in such cases unfavourable, electron addition equilibrium to the right.

The Birch reduction:

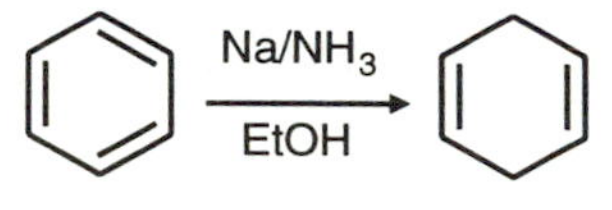

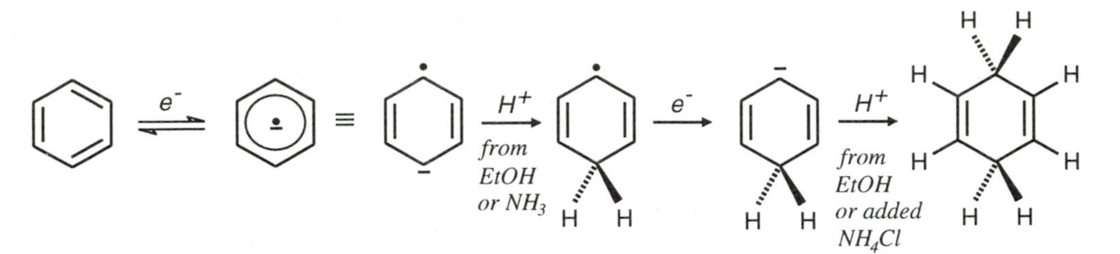

Some substituents (COOH, $CONH_2$, OR, NR_2 and alkyl groups) survive the reaction conditions, but the orientation of the reduction depends on the electronic nature of any groups attached to the ring. In the 1,4-dihydrobenzene (cyclohexa-1,4-diene) product, electron-withdrawing substituents are found on a saturated carbon atom whereas electron-donating substituents are located on a double bond:

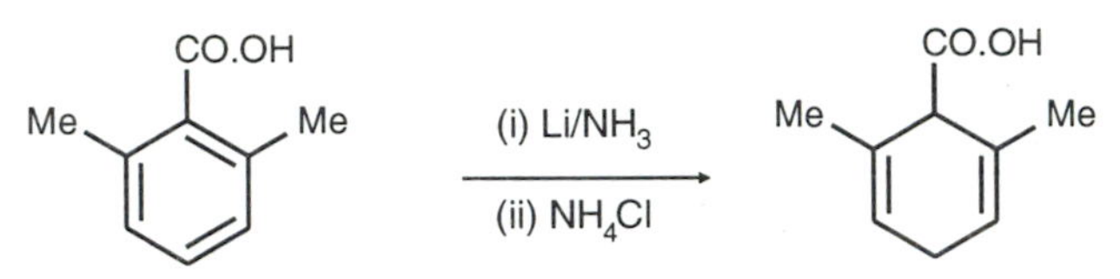

Of particularly synthetic importance is the reduction of anisole and *N,N*-dialkylanilines, since, in each case, the intermediate dihydroaromatic product is readily hydrolysed to a cyclohexenone.

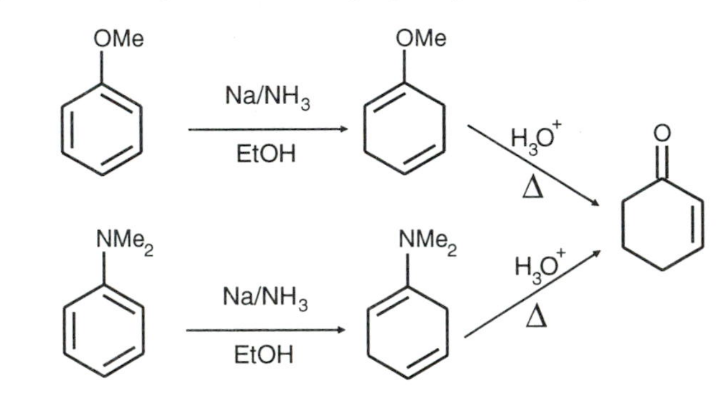

Birch reduction and catalytic hydrogenation frequently produce opposite stereo-chemical results (catalytic hydrogenation often shows marked solvent sensitivity):

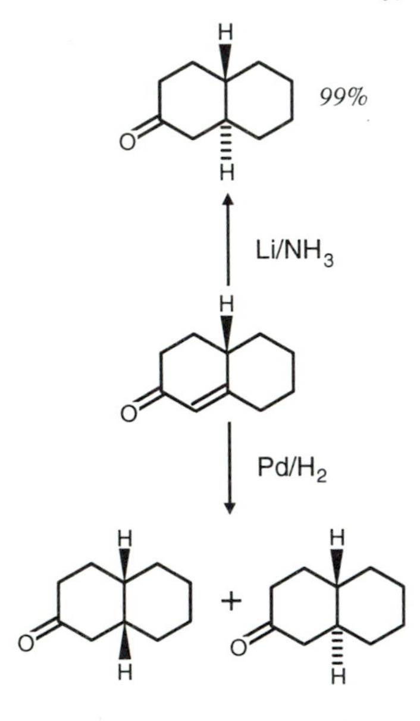

3.9 Catalytic hydrogenation of arenes

Catalytic hydrogenation of aromatic rings offers a useful route to saturated carbocyclic rings. Reduction of benzene derivatives provides a route to substituted cyclohexanes, hydrogen addition normally occurring from the less-hindered side of the molecule.

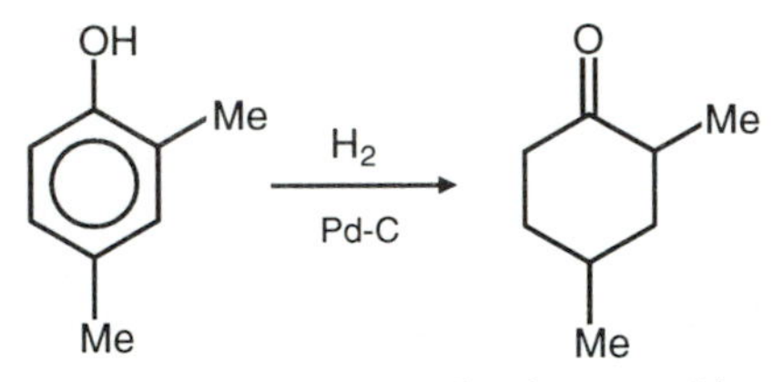

Polynuclear aromatic compounds also provide a useful source of fused polycyclic structures, *e.g.* tetralin (tetrahydronaphthalene) and decalin derivatives are readily obtained from the corresponding naphthalenes.

4: Conformation and reactivity in alicyclic compounds

4.1 Effect of substituent stereochemistry on reactivity of cyclohexane derivatives

As was shown earlier, a substituent on a cyclohexane ring can occupy an axial or an equatorial location, the latter being the generally favoured situation. However, the presence of other groups can perturb this conformational equilibrium and, if the other group is very bulky, this can lead to "conformational locking" of the ring geometry *e.g.* by a *tert*-butyl substituent. Under such circumstances it is possible to probe the effect of different conformations on the reactivity of substituted cyclohexanes. It is convenient to consider two main types of effect here, namely steric and stereo-electronic factors.

4.2 Steric Effects

Steric effects, *i.e.* van der Waals forces, derive from large and repulsive close contacts, or small attractive interactions operating over greater distances. Repulsive interactions are the most readily identified and can operate in both ground and transition states. When steric interference is significant in the transition state but of little impact in the ground state there will be an increase in the activation energy for the process and the reaction rate may be said to be retarded by "steric hindrance". On the other hand, if the ground state is significantly more destabilised by steric interference than the transition state, the activation energy will be reduced leading to an accelerated reaction rate because of "steric assistance".

Increased steric repulsion in a reaction transition state slows the rate of reaction as a result of "steric hindrance":

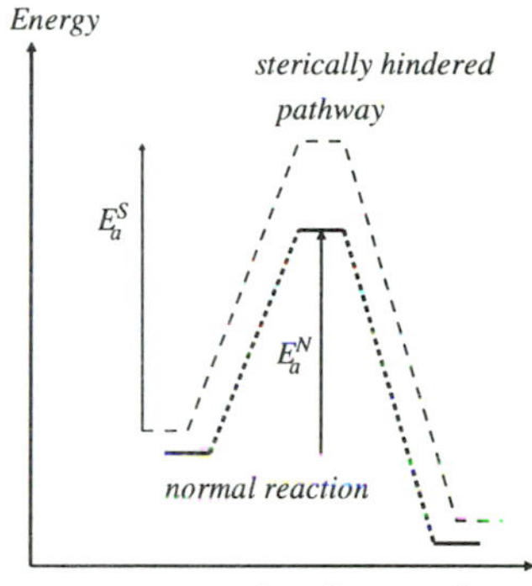

$k^N > k^S$ since $E_a^S > E_a^N$

4.2.1 Relative rates of ester hydrolysis The hydrolysis of ethyl cyclohexanone carboxylates provides an example of the consequences of steric hindrance. Comparison of the rate of reaction of this ester with those of its *cis*- and *trans*-4-*tert*-butyl derivatives reveals the following results:

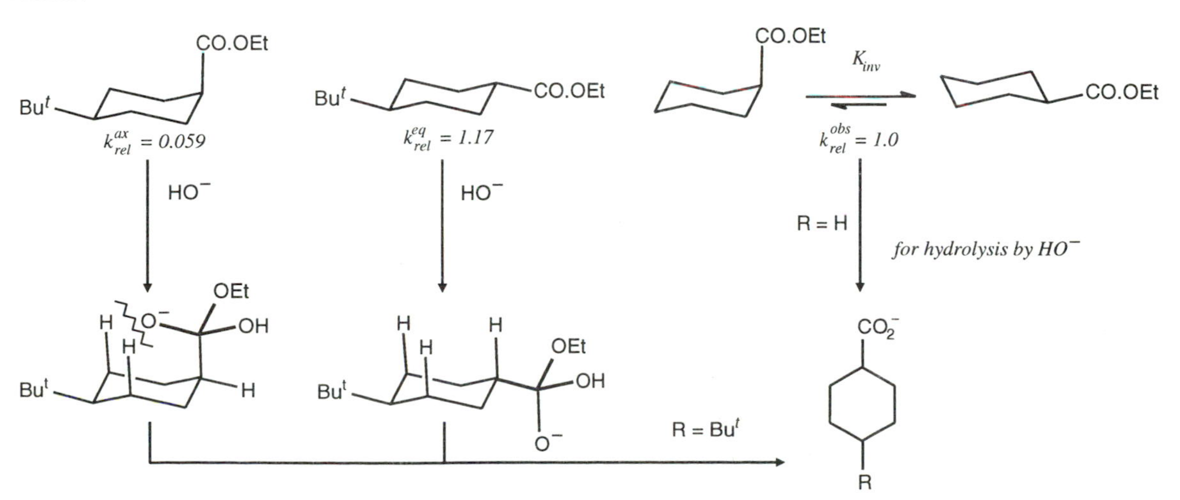

From the rate constants it is possible to deduce a value for K_{inv}, the equilibrium constant for the conformational inversion of ethyl cyclohexane carboxylate. If it is assumed that the rates of hydrolysis of the axial and equatorial esters are the same as their conformationally locked analogues, then:

$$k_{rel}^{obs} = k_{rel}^{ax} \times N^{ax} + k_{rel}^{eq} \times N^{eq}$$

where N^{ax} and N^{eq} are the mole fractions of the axial and equatorial isomers (*i.e.* $N^{ax} + N^{eq} = 1$). This gives a value for N^{eq} of 0.85 and $K_{inv} = 5.54$ ($\Delta G_{inv} = -4.2$ kJ mol^{-1} at 298K).

During the rate-determining transition state for the hydrolysis process a hydroxyl group becomes attached to the carbonyl carbon atom which therefore changes from sp^2 to sp^3 hybridisation. This increases the bulk of this substituent (whereas the planar ester group can readily reduce crowding by rotation). Consequently the activation energy for hydrolysis of the axial ester is significantly greater than that for its equatorial counterpart leading to a *ca.* 20-fold difference in reaction rates. [This discussion ignores the fact that the transition state is negatively charged whereas the ground state is neutral leading to significant differences in solvation between the two which also favour the equatorial isomer.] The rate of hydrolysis of the conformationally mobile unsubstituted ester is intermediate between those of the two 4-*tert*-butyl derivatives since this value is the mole fraction-weighted average of those for pure axial and pure equatorial esters.

4.2.2 Steric assistance in the oxidation of cyclohexanols An example of steric assistance is provided by the relative rates of oxidation of the anomeric 4-*tert*-butylcyclohexanols. In this case the axial *cis*-alcohol is oxidised by Cr^{VI} *ca.* 3.2 times faster than its equatorial *trans*-isomer. The rate determining step is fragmentation of the bulky intermediate chromate ester as shown by a kinetic isotope effect of *ca.* 6 [a value which indicates significant C-H (or C-D) bond cleavage in the rate determining step]. In the axial isomer, this process is accompanied by strain relief and it therefore reacts faster.

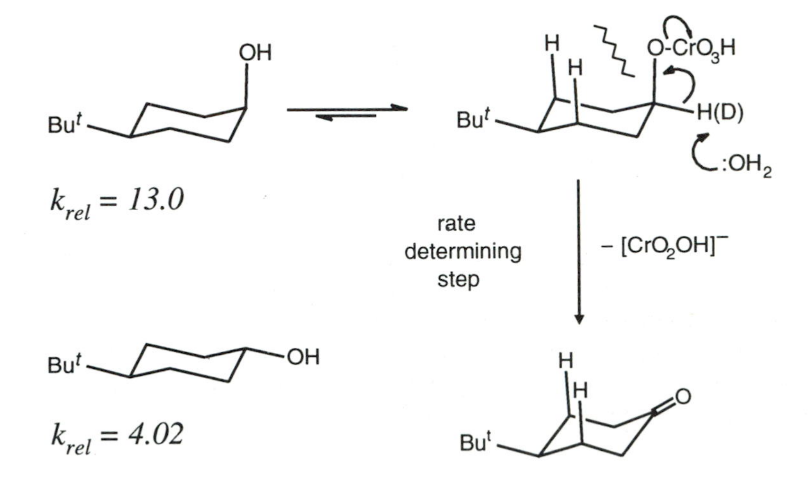

Similar results are seen in steroidal systems. Here the rate enhancement brought about by relief of steric compression during the oxidation of rigid steroidal alcohols falls off rapidly as the distance between the interacting groups increases since this reduces the steric interactions:

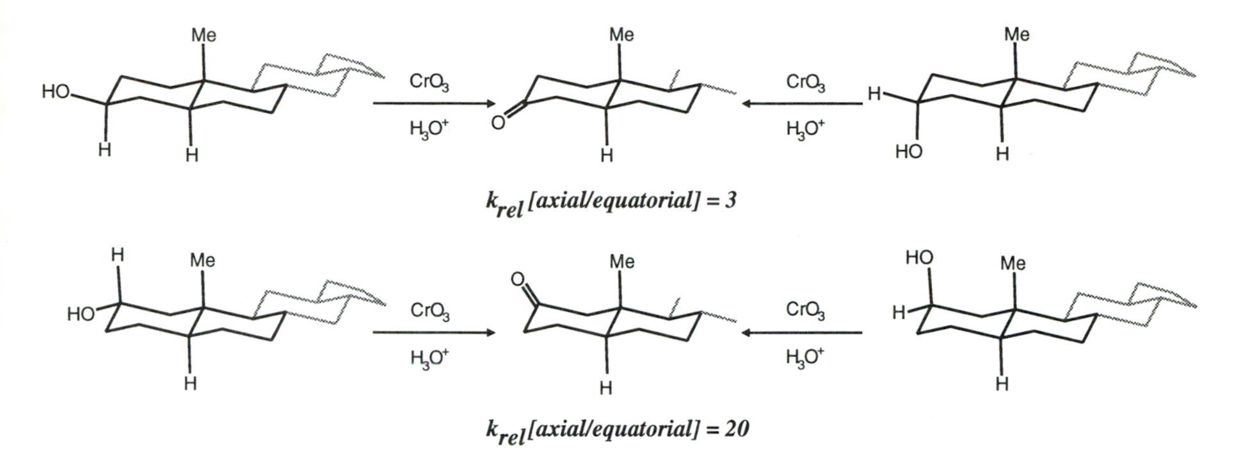

4.3 Stereoelectronic effects

Deslongschamps defines such effects as changes in reactivity arising from the spatial disposition of particular bonded or non-bonded electron pairs. For example, there is a stereoelectronic requirement for an S_N2 reaction to proceed with inversion of configuration which arises because the new σ-bond is formed by attack of the nucleophile on the σ*-orbital of the bond which is breaking. Nucleophilic substitution in epimeric alkyl halides proceeds at different rates:

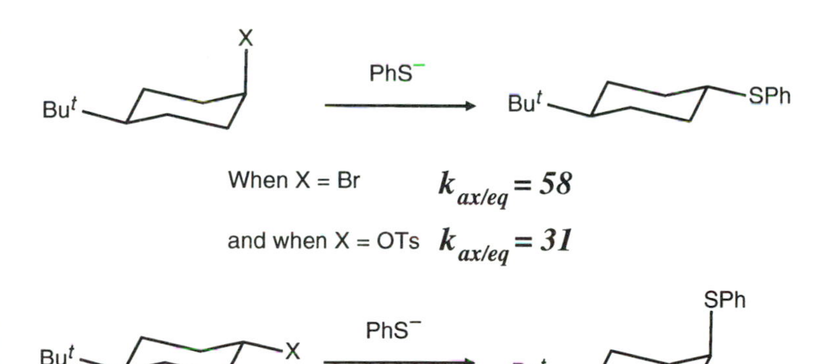

In general the axial isomer reacts faster than its equatorial equivalent, a result which reflects a combination of steric and stereoelectronic factors. In particular these include relief of ground-state compression (a steric effect) in the axial case, and steric hindrance to backside attack in the equatorial derivative (for which the stereoelectronic requirement is attack from an axial direction).

Substitution of an equatorial substituent by an S_N2 pathway requires backside attack by the nucleophile, a process which is sterically hindered by axial substituents around the ring:

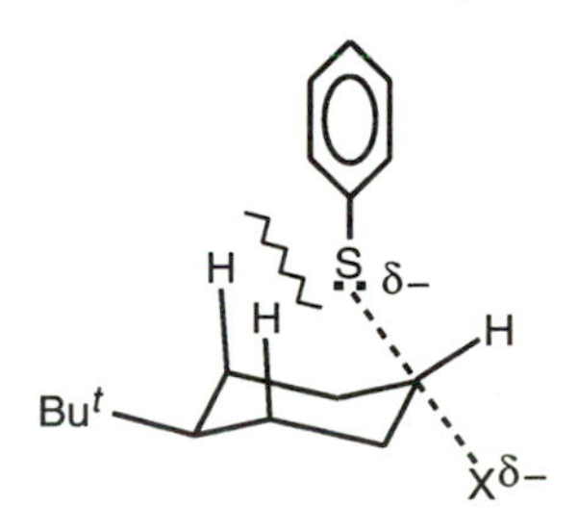

4.3.1 Addition and elimination Another reaction pathway which has a stereoelectronic requirement is bimolecular elimination (E2). This process occurs most readily when the eliminated groups are located antiperiplanar with respect to each other. This leads to optimal alignment of the σ-bond being attacked by the base and the σ*-orbital of the bond attached to the leaving group, and requires minimum structural reorganisation to form the new π-bond.

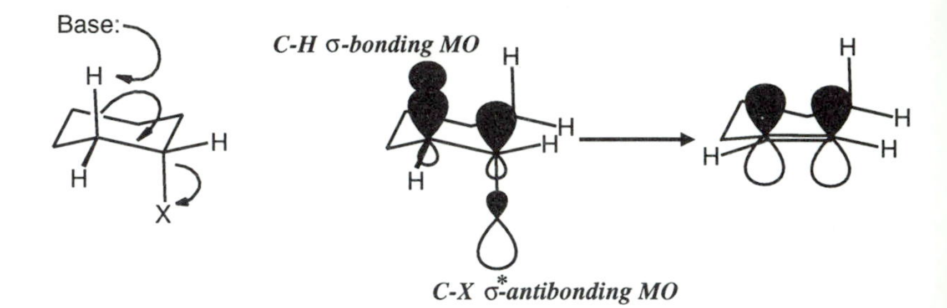

The next best orientation from the point of view of optimal orbital overlap is a synperiplanar one, but this does not occur in a chair cyclohexane. Eliminations involving equatorial bonds involve a synclinal orientation which is unfavourable. Accordingly it appears that E2 eliminations in cyclohexanoid systems will occur using neighbouring *trans*-diaxial groups but not *trans*-diequatorial or *cis*-axial-equatorial substituents. So, for example, whilst *cis*-4-*tert*-butylcyclohexyl tosylate readily eliminates (by an E2 mechanism) the *trans* isomer, in which the tosylate group is equatorial, prefers E1 (and competing S_N) reaction pathways. Cyclohexyl tosylate itself undergoes E2 elimination at a rate about 0.26 that of the *cis*-4-*tert*-butyl derivative, reflecting the preference for a conformation in which the tosylate group is equatorial.

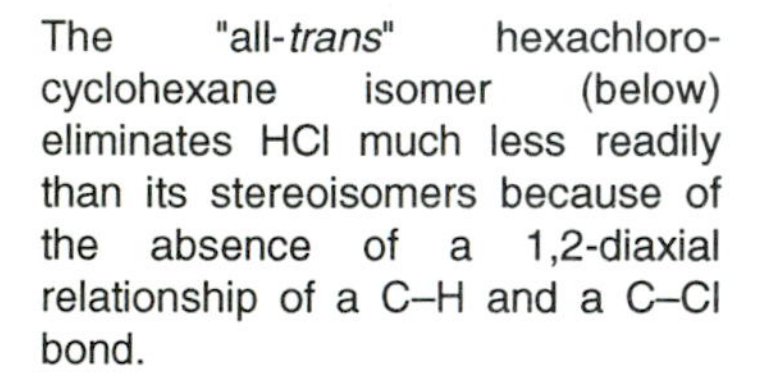

The "all-*trans*" hexachlorocyclohexane isomer (below) eliminates HCl much less readily than its stereoisomers because of the absence of a 1,2-diaxial relationship of a C–H and a C–Cl bond.

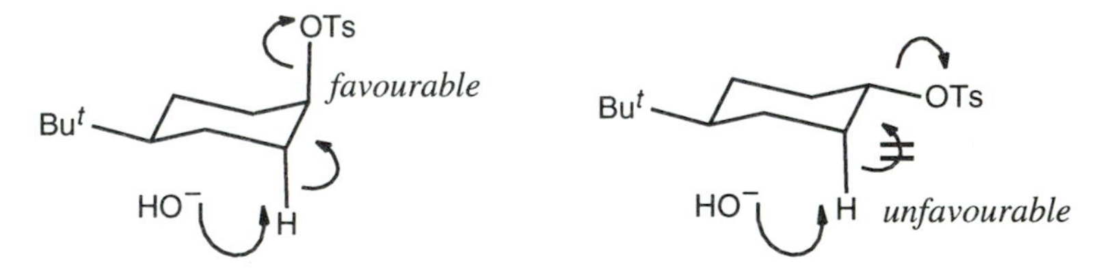

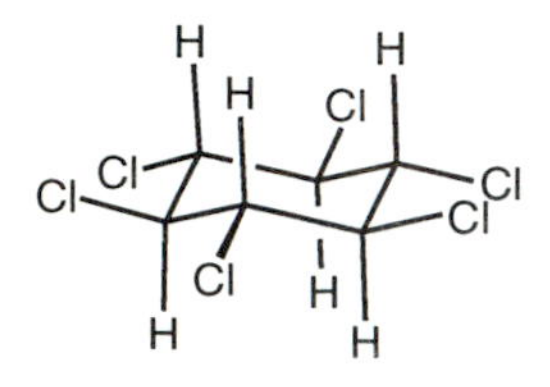

The nine stereoisomers of 1,2,3,4,5,6-hexachlorocyclohexane include seven *meso* forms and a *d,l* pair. Only one of these (shown left) has no chlorine substituent antiperiplanar *(i.e.* in a 1,2-*trans* diaxial relationship) with respect to a hydrogen atom in its preferred conformation. Consequently this isomer undergoes base-catalysed elimination 7000 times slower than any of the other isomers.

Another comparison of the relative ease of such diaxial and axial-equatorial eliminations is provided by the iodide-induced debromination of *cis*- and *trans*-1,2-dibromocyclohexane. Formation of cyclohexene from the *trans*-isomer is *ca.* 11 times faster than from the *cis*, but this does not give direct insight into the relative ease of these two processes since kinetic studies suggest that the *cis*-isomer undergoes a rate-determining substitution of bromide by iodide prior to elimination. Account must also be taken of the position of the diequatorial-diaxial conformational equilibrium for the *trans*-dibromo isomer. From this it is clear that diaxial elimination occurs faster than bimolecular substitution.

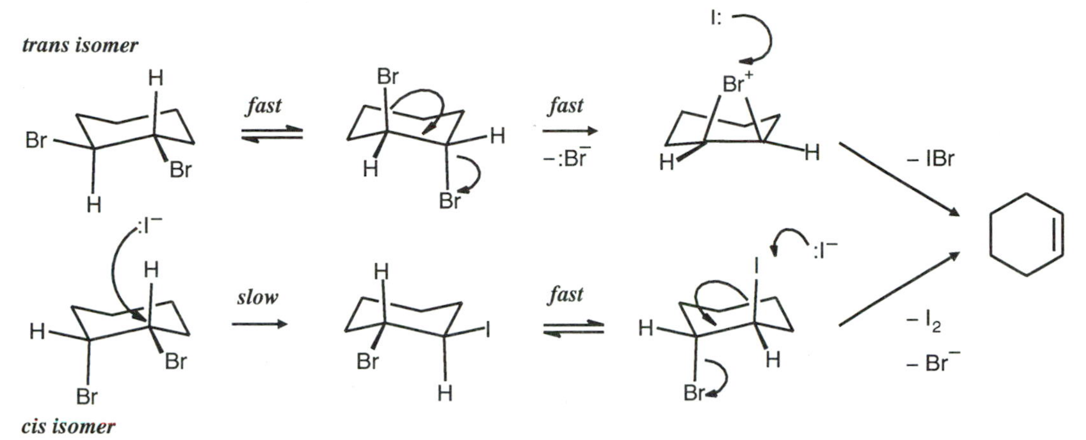

The addition of bromine to cyclohexene follows the reverse of this pathway preferring to form the *trans*-antiperiplanar product through nucleophilic attack on an intermediate bromonium ion. The alternative site of attack leads to a less-favourable diaxial twist-boat geometry. Each of these must then ring-invert into the corresponding 1,2-diequatorial chair form:

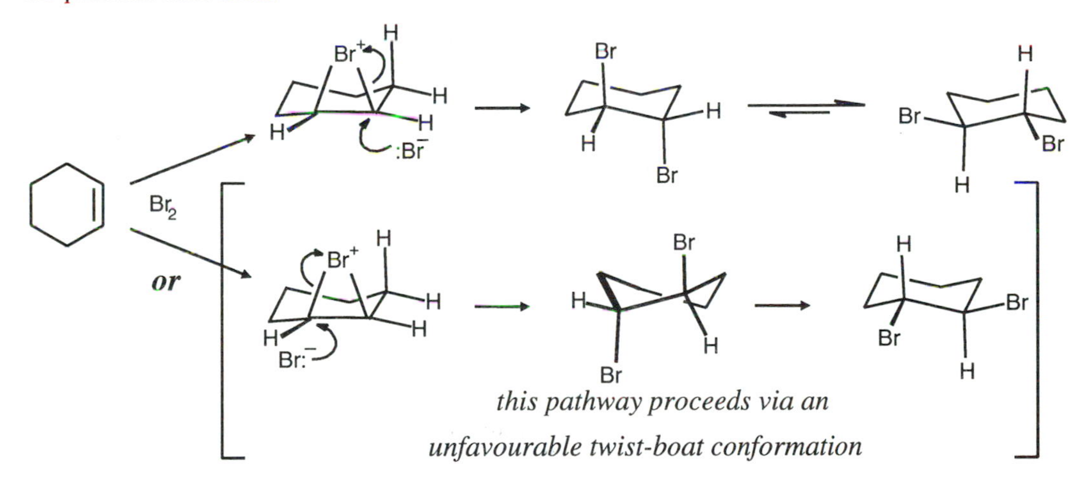

When bromination of cyclohexene is carried out in water the result is the formation of a *trans*-1,2-bromohydrin, formed once again by diaxial opening of an intermediate bromonium ion.

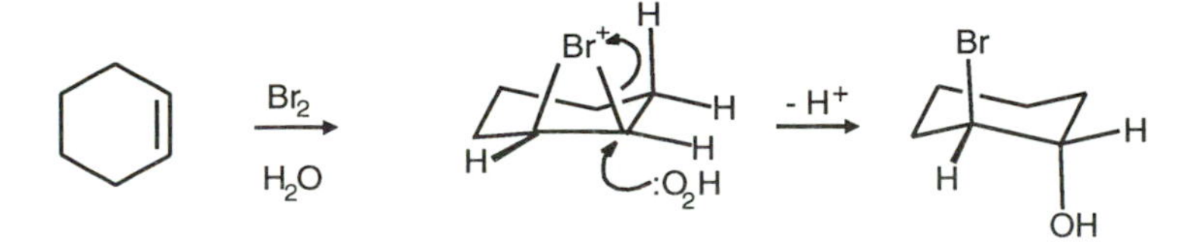

The reactions of conformationally locked bromohydrins in the presence of base or Ag_2O provide another nice illustration of stereoelectronic control; in each case the preferred pathway is determined by the nature of the substituent/bond antiperiplanar to the leaving group, a bromide ion.

When the hydroxyl and bromo substituents are both axial, an internal nucleophilic displacement leads to epoxide formation:

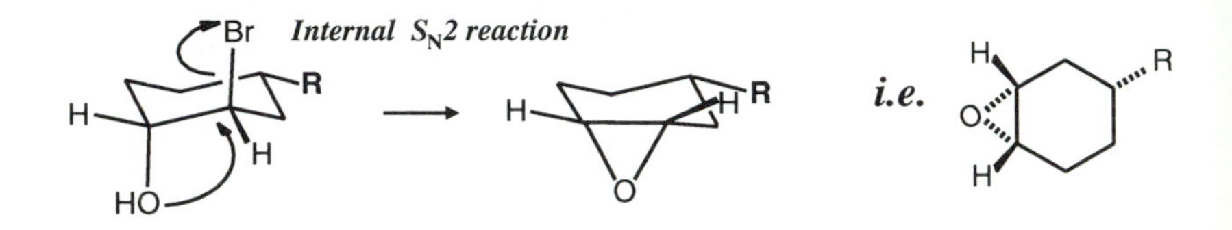

whereas with bromine axial and the hydroxyl group equatorial, the favoured pathway involves a hydride shift (or HBr elimination) of the axial C-H bond.

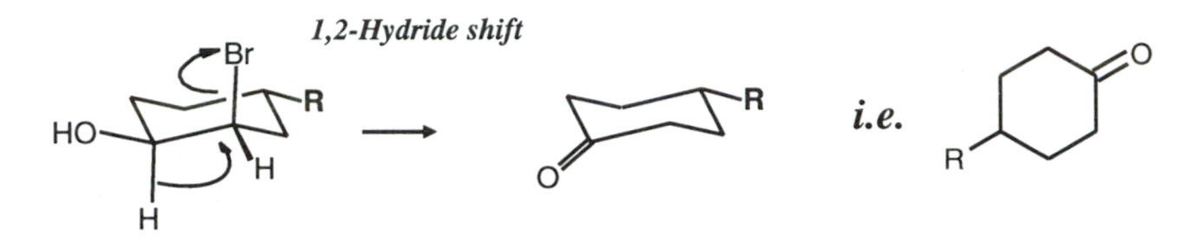

If both substituents are equatorial, ring contraction is possible involving a 1,2- (Wagner-Meerwein) shift of the C-C bond *trans coplanar* with the C-Br bond. This leads to a hydroxyl-stabilised carbocation which loses a proton resulting in formation of a carbonyl group. Further examples of such skeletal rearrangements are discussed later in section 4.3.

However, if the conformational lock is not too bulky, ring inversion can occur to give the diaxial substituent geometry required for epoxide formation (but with inverted stereochemistry relative to the example above).

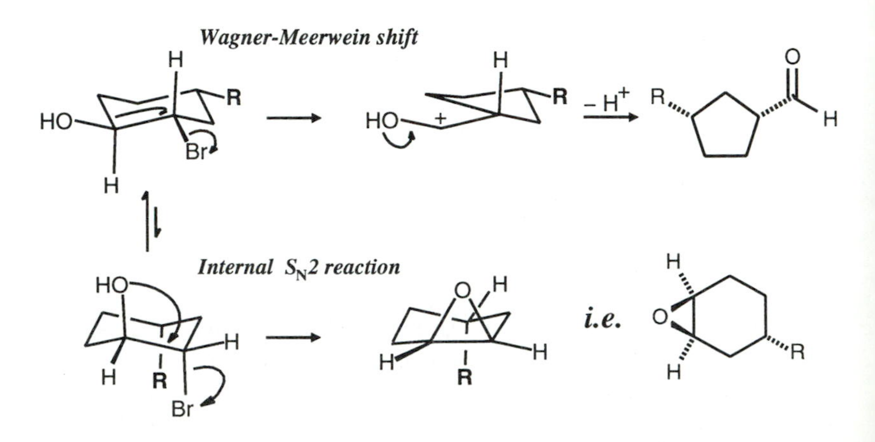

4.3.2 Elimination reactions in other alicyclic rings We have already noted that whilst an antiperiplanar geometry gives optimum orbital alignment for an elimination process, a syn-periplanar arrangement of departing groups is almost as good. Such an elimination process becomes important when an antiperiplanar conformation cannot be achieved. For example, elimination of H(D)–X from norborn-2-yl derivatives such as that shown below (which are effectively rigid, bridged cyclohexane boat conformations) affords a norbornene which has no deuterium present, reflecting the favourable syn (0° for D-C-C-X) and unfavourable anti (120° for H-C-C-X) dihedral angles for elimination.

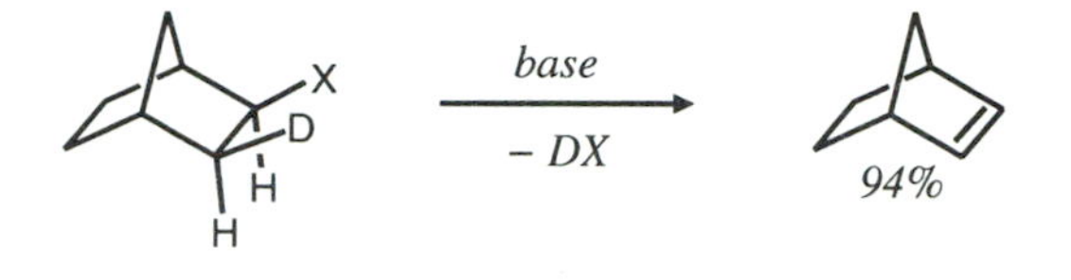

Six-membered rings are the only alicyclic structures between C_4 and C_{14} in which a strain-free antiperiplanar conformation can be readily achieved. Consequently syn elimination becomes more important for other ring sizes and is frequently the predominant process, particularly in cycloalkyltrimethylammonium hydroxides (Hofmann elimination):

Hofmann elimination occurs when an alkyltrimethyl ammonium hydroxide is heated:

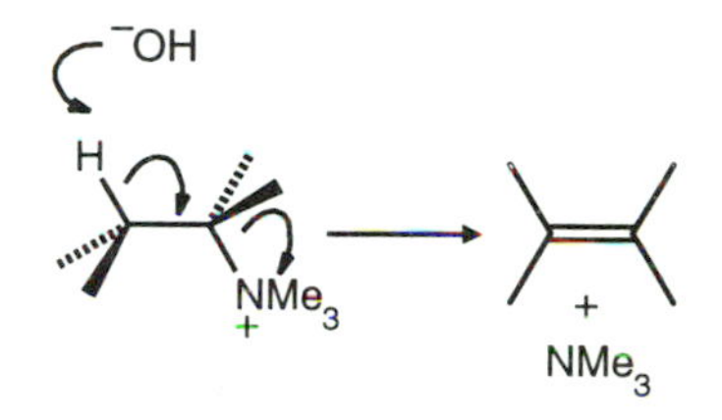

% Syn Elimination in Cycloalkyltrimethylammonium hydroxides				
Ring size	4	5	6	7
% Syn product	90	46	4	31-37

For medium ring derivatives the possibility of *cis* or *trans* isomer formation introduces a further complication. Whilst the *cis* isomer is more stable, Hofmann elimination gives almost pure *trans*-cycloalkene from both cyclononyl- and cyclodecyl trimethylammonium hydroxide.

One explanation for these observations is that in its preferred conformation, the cyclodecyl salt prefers the conformation shown below in which anti-coplanar elimination leads directly to *trans*-cyclodecene.

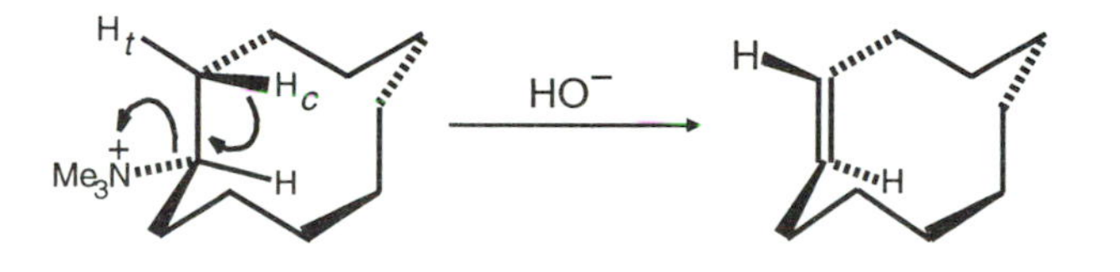

In the cyclodecyl case Hofmann elimination involves the *anti*-elimination of groups disposed *cis* about a cyclodecane ring rather than the seemingly more likely process of *syn*-elimination in the conformation shown below:

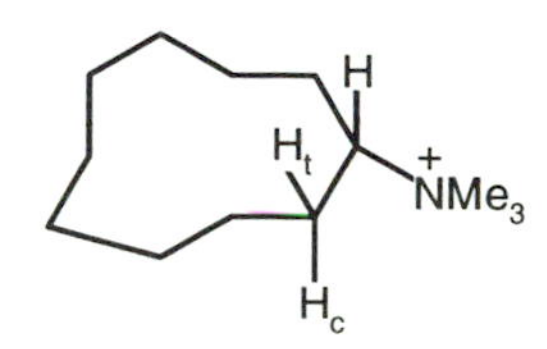

Selective isotopic labelling has provided further information about this process. A substantial isotope effect (indicative of an E2 mechanism) is observed for the formation of both *cis* and *trans* cyclodecenes only when H_t is deuterated (in 1,1,4,4-tetramethyl-7-

cyclodecyl trimethylammonium hydroxide). This suggests that *only* the *trans* hydrogen is lost when either cycloalkene isomer is formed, *i.e.* that the *cis* isomer results from anti-elimination whereas the *trans* isomer is formed by syn-elimination. This interpretation, which is known as the "syn-anti dichotomy", has however been questioned.

Even elimination in cyclo-octyl trimethylammonium hydroxide favours the formation of *trans*-cyclo-octene (*trans*:*cis* = 3:2) despite the fact that this is thermodynamically much less stable than its *cis* isomer .

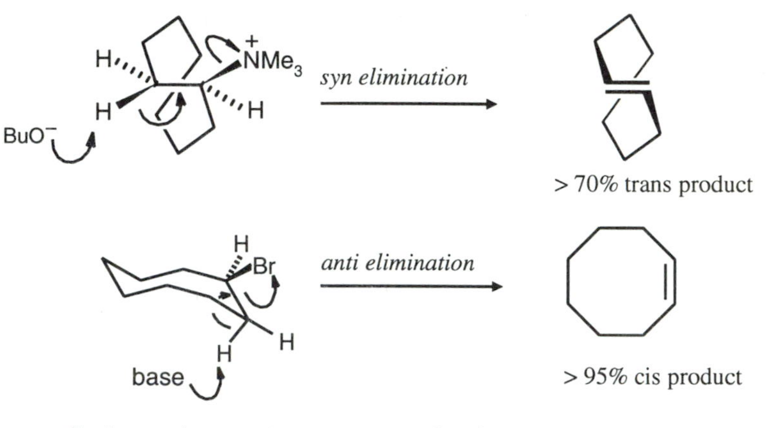

Such product ratios are very leaving-group dependent. For example, dehydrobromination leads predominantly to the *cis* isomer suggesting that leaving group bulk may be an important factor in determining the favoured reaction pathway.

4.3.3 Elimination reactions forming cycloalkynes

Dehydrohalogenation of cycloalkenyl halides does produce cycloalkynes (though allenes can also be formed). Even cyclohexyne can be formed in this manner, for example by reaction of 1–bromocyclohexene with PhLi, but is immediately trapped by this reagent. When a weaker base (*e.g.* *t*-BuOK) is used the allene dimer is obtained as product. However, it is not certain whether the allene intermediate is formed directly or arises from base-catalysed alkyne-allene isomerisation.

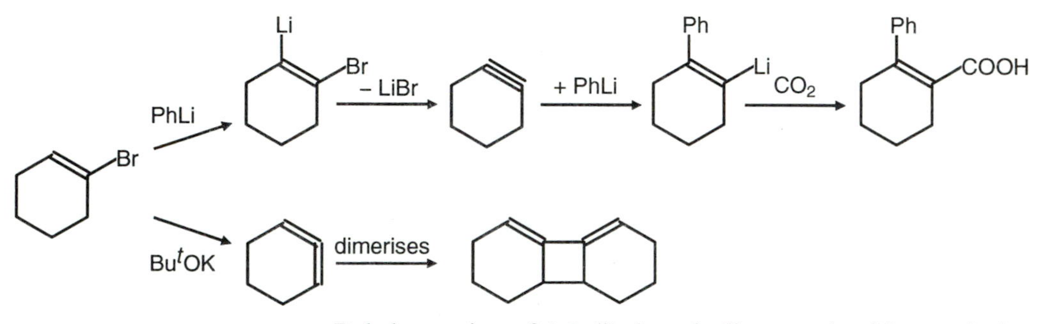

Dehalogenation of 1,2-dihalocycloalkenes using Na metal gives cycloalkynes in poor yield.

4.4 The chemistry of cyclohexene epoxides

Reaction of cyclohexene with peracids leads to formation of the corresponding epoxide which, like its alkene precursor, exists in a half chair conformation. Whilst cyclohexene oxide must, of necessity, have *cis*-stereochemistry, *trans*-isomers are also known for larger rings.

4.4.1 Face-selective epoxidation In the presence of a substituent, the faces of a cycloalkene double bond are no longer equivalent. Epoxidation of 4-*tert*-butylcyclohexene is almost random (*cis*:*trans* = 3:2) but for the 3-*tert*-butyl isomer formation of the *trans*-isomer is favoured (by a factor of 9:1). It would seem that, as for bromination, a pseudoequatorial bulky substituent blocks approach of the electrophile from the face *cis* to it.

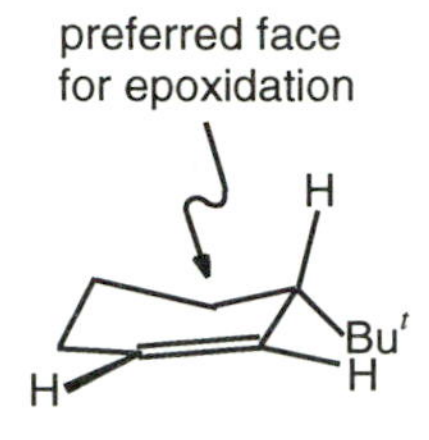

Other interactions can also influence the facial selectivity of epoxidation. For example, when there is an allylic hydroxyl group present hydrogen-bonding in the transition state leads to almost complete *cis*-selectivity (which is destroyed when the OH substituent is acylated).

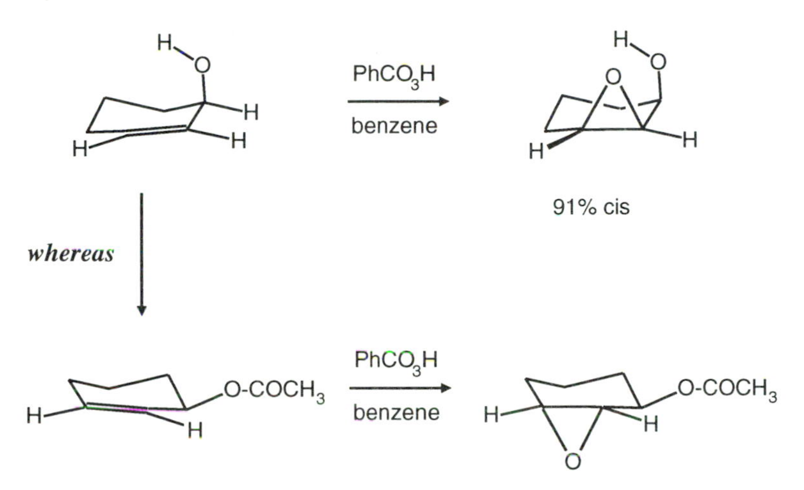

Intermolecular hydrogen bonding can influence the stereoselectivity of epoxidation

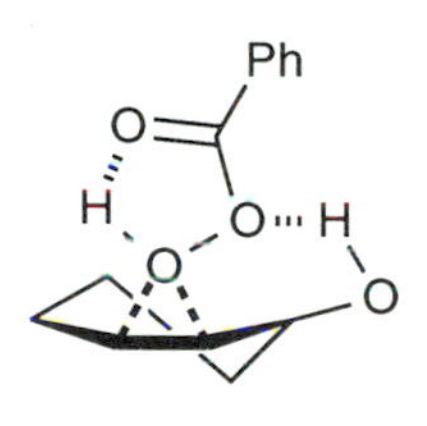

The stereochemical selectivity of transition metal-catalysed epoxidations can be strongly influenced by the presence of co-ordination sites within the molecule. Molybdenum-catalysed epoxidation (the Sharpless reaction) of the homoallylic cyclohexen-4-ol again shows almost exclusively *cis*-selectivity. Similar results are obtained when titanium and vanadium catalysts are employed.

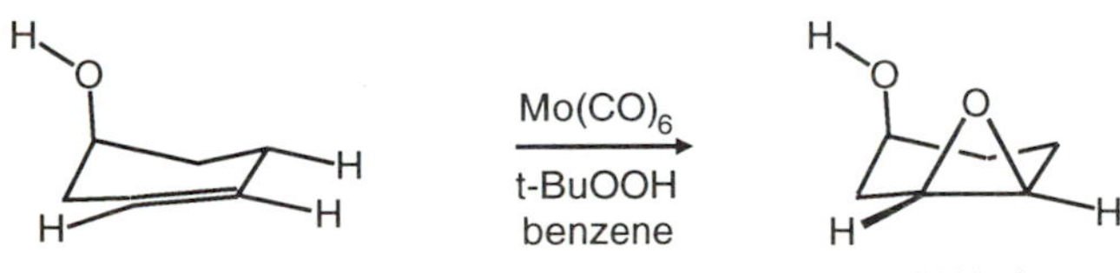

Metal-ion co-ordination can also lead to stereoselectivity in epoxidation reactions:

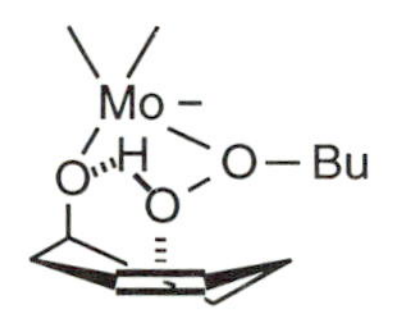

4.4.2 Stereochemistry of epoxide ring opening As with bromination, ring opening of cyclohexene epoxides generally leads to the diaxial rather than the diequatorial product. This is illustrated by the lithium aluminium deuteride-induced opening of *cis*-4-*tert*-butylcyclohexene epoxide shown below.

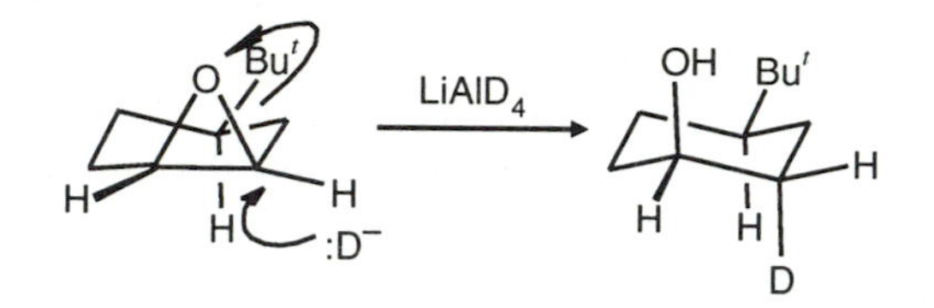

When the cyclohexene ring is part of a *trans*-decalin skeleton, the presence of an axial bridgehead methyl substituent shields the *syn*-face from electrophilic attack. Direct epoxidation by a peracid proceeds on the opposite face. The opposite epoxide stereochemistry is achieved *via* a bromohydrin. In this case the intermediate bromonium ion is also formed preferentially on the *anti*-face and nucleophilic ring opening then occurs in a *trans*-diaxial manner (see section 4.3.1) to give the *trans*-diaxial bromohydrin:

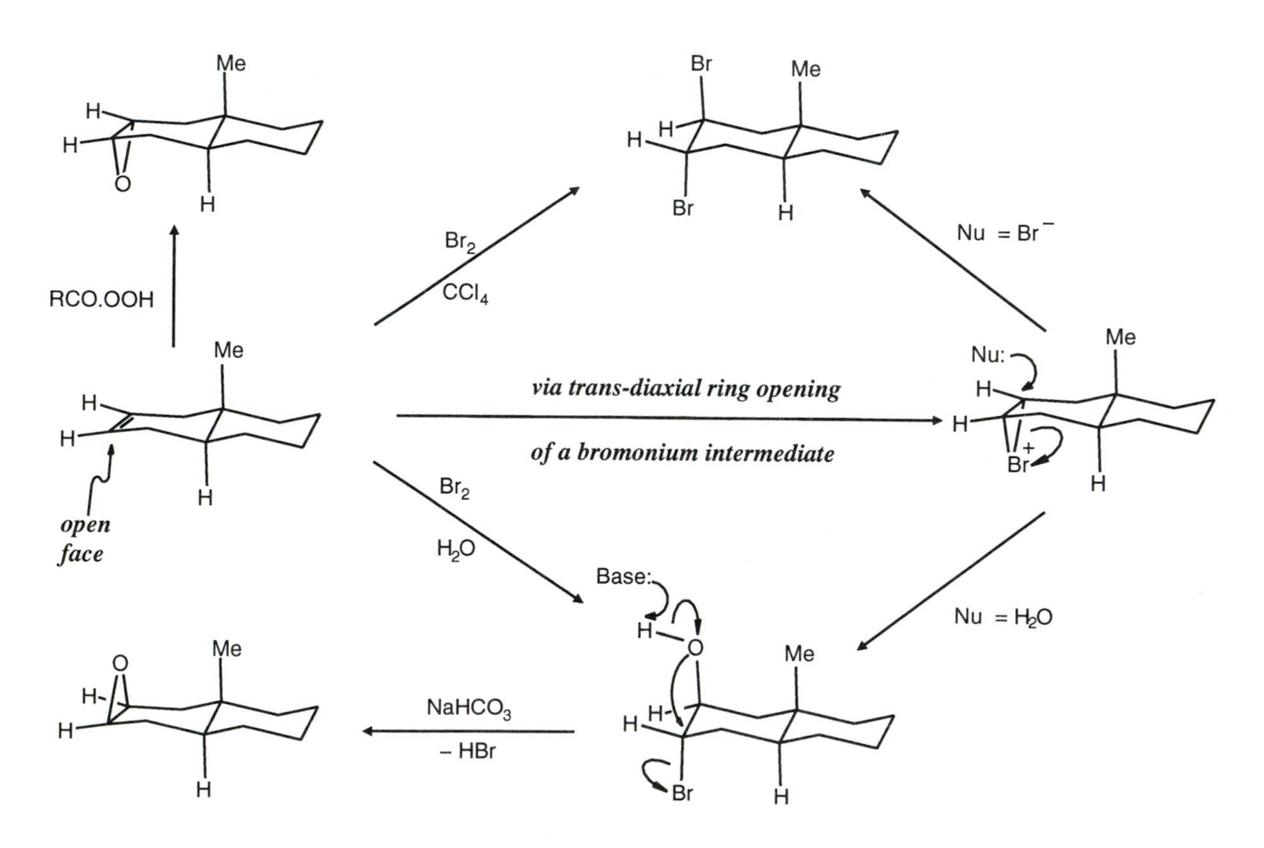

When this bromohydrin is treated with base, internal nucleophilic displacement occurs with loss of HBr to give the *cis*-epoxide.

4.5 Reduction of ketones

Hydride reduction of cyclohexanones represents a controversial problem. Reaction of cyclohexanones with lithium aluminium hydride and its analogues can be regarded as nucleophilic addition of hydride to a carbonyl group. Reduction of 4-*tert*-butylcyclohexanone with $LiAlH_4$ gives a product mixture in which the more stable equatorial alcohol predominates (see Table) - a result which is typical for "small" reducing agents. However, when bulky hydride reducing agents such as (L)-selectride [LiHB(sec-Bu)$_3$] are used, the product is almost exclusively the axial alcohol *i.e.* the *cis*-isomer.

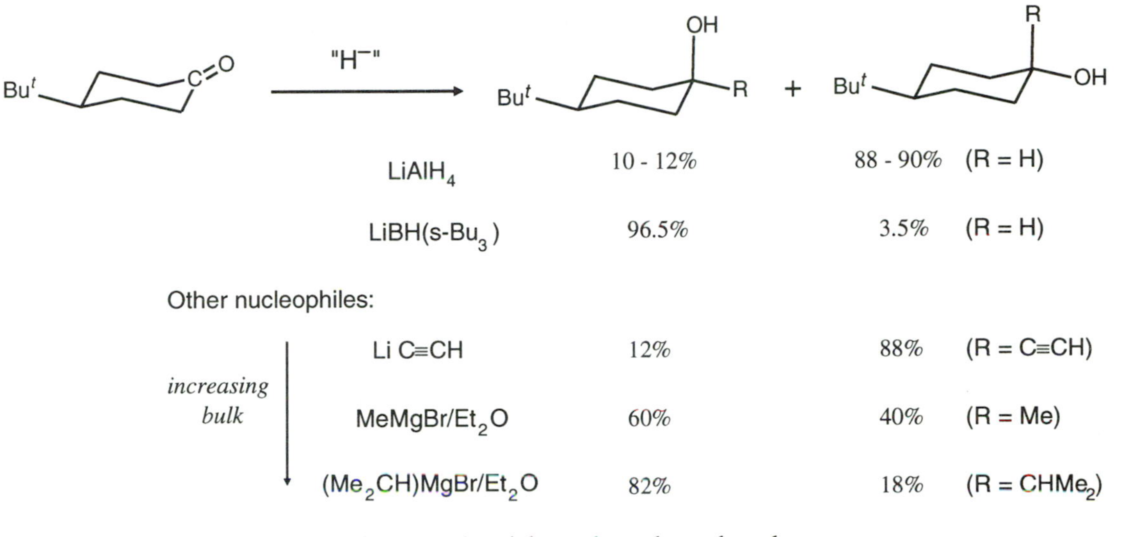

$LiAlH_4$	10 - 12%	88 - 90%	(R = H)
LiBH(s-Bu$_3$)	96.5%	3.5%	(R = H)
Other nucleophiles:			
Li C≡CH	12%	88%	(R = C≡CH)
$MeMgBr/Et_2O$	60%	40%	(R = Me)
$(Me_2CH)MgBr/Et_2O$	82%	18%	(R = $CHMe_2$)

increasing bulk ↓

These results suggest that, in general, axial attack on the carbonyl group is favoured by nucleophiles, leading to equatorial alcohol product. However, axial addition may be hampered by steric factors resulting in an increased preference for equatorial attack (favouring axial alcohol product) as shown by the change in product distribution as the bulk of the nucleophile is increased (see Table).

Substituents about the cyclohexane ring can also influence the stereochemistry of nucleophilic addition to the carbonyl group. $LiAlH_4$ reduction of 2-methylcyclohexanone favours axial-hydride addition whereas when the more bulky (L)-selectride [LiBH(*s*-Bu$_3$)] is used only equatorial addition of hydride is seen.

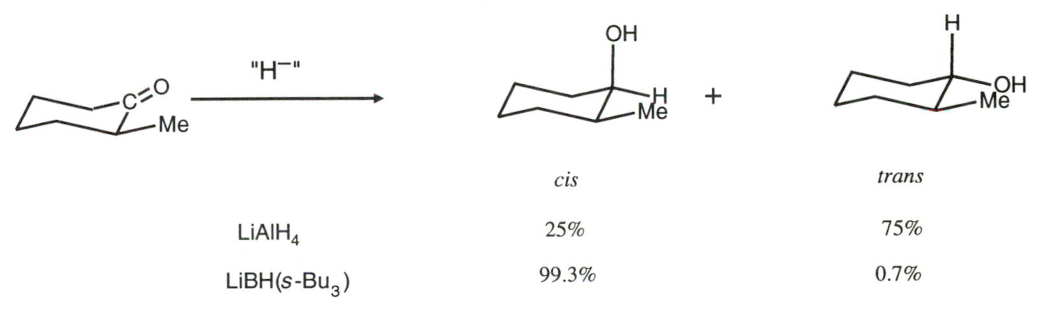

	cis	*trans*
$LiAlH_4$	25%	75%
LiBH(*s*-Bu$_3$)	99.3%	0.7%

$LiAlH_4$ reduction of a 3,5,5-trimethylcyclohexanone is much less selective because the axial methyl group interferes with attack of nucleophile from this direction.

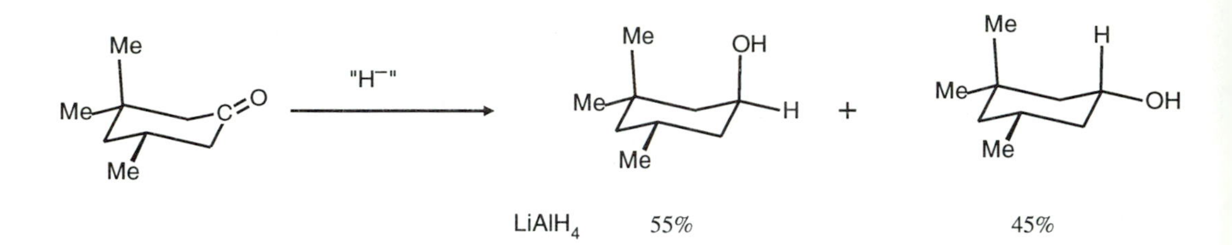

The reductions discussed above are kinetically controlled whereas sodium in ethanol converts 4-*tert*-butylcyclohexanone almost exclusively (~98%) into the *trans* (*i.e.* equatorial) alcohol and use of aluminium isopropoxide (Meerwein-Ponndorf-Verley reduction - an equilibrium controlled process) shows that, at equilibrium, the equatorial alcohol predominates ($K_{eq/ax}$ = 3.76 at 365K).

The Meerwein Ponndorf Verley reduction is an equilibrium-controlled process and consequently the proportions of the different products reflect their relative stabilities.

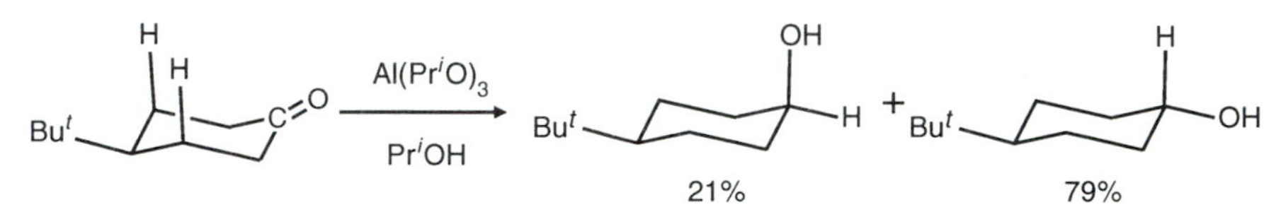

Steric hindrance of nucleophilic addition to carbonyl groups becomes even more important in bi- and polycyclic structures. For example, hydride reduction of norbornanone leads predominantly to *endo*-alcohol reflecting the far more open nature of the *exo*-face of this structure (*n.b.* an equilibrium mixture of the two contains *ca.* 90% of the *exo*-alcohol so hydride-reduction is not a thermodynamically controlled process). However, in camphor there is a methyl group at C_7 which shields the *exo*-face from nucleophilic addition and *exo*-alcohol (derived from hydride attack on the *endo* face) now predominates.

A *syn*-7-methyl group hinders attack from the normally less hindered *exo*-face of a norbornan-2-one derivative.

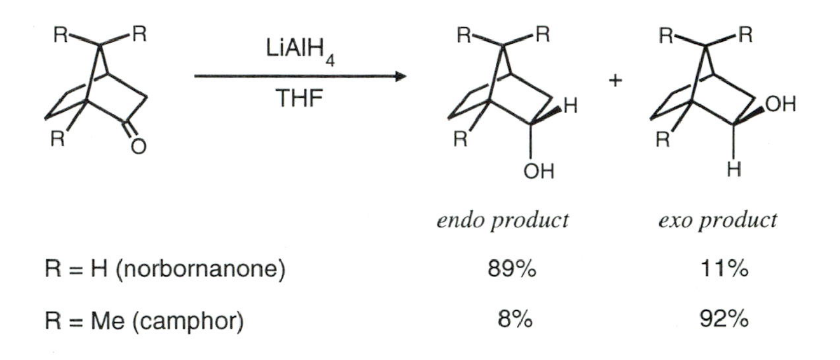

4.5.1 Cyanohydrin formation In the absence of other factors there is a general preference for axial addition to cyclohexanones. For example, both 2- and 3-methyl cyclohexanone form cyanohydrin acetates which result from predominantly axial attack by the nucleophile (CN^-).

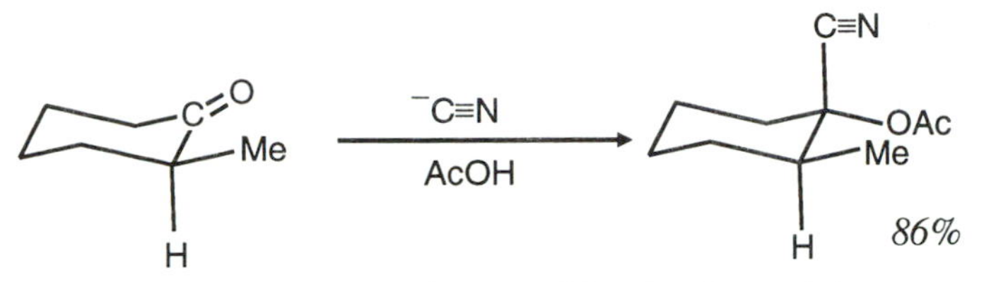

4.5.2 Why is axial addition favoured in simple cyclohexanones? A number of theories have been advanced to explain this observation. Felkin has proposed an explanation based on steric interactions in an early, reactant-like transition state:

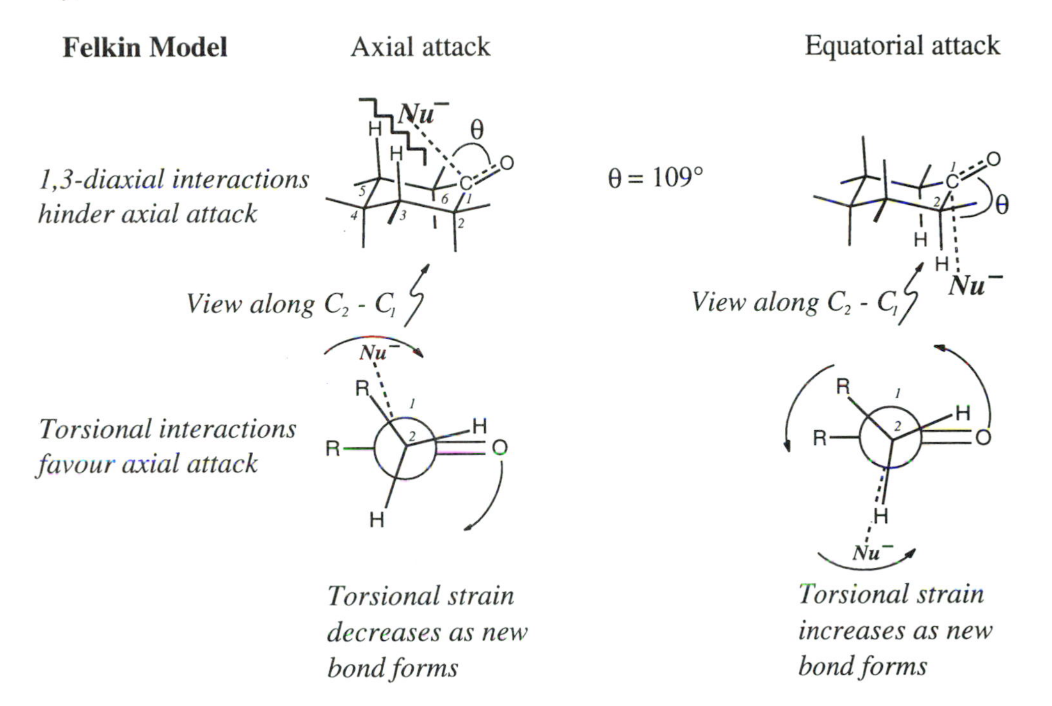

It is perhaps not surprising that bulky groups should favour attack from the equatorial direction in order to avoid unfavourable 1,3-diaxial interactions with substituents at C_3 and C_5. However, when addition from an equatorial direction is viewed along the C_2–C_1(=O) bond it is seen that, as the carbonyl carbon atom becomes tetrahedral, the C–O bond sweeps into an axial position passing through unfavourable eclipsing interactions with neighbouring C–H bonds on C_2 and C_6. Attack from the axial face by small nucleophiles is much less affected by 1,3-diaxial interactions and rehybridisation of the carbonyl group results in the C–O bond twisting in the opposite direction, avoiding such unfavourable torsional clashes.

An electronic explanation has also been proposed which suggests that during hydride reduction, a stabilising interaction occurs involving the σ* molecular orbital of the developing C-H bond (between the hydride moiety and the carbonyl carbon atom) and the σ-bonding molecular orbitals of the axial C-H groups at C_2 and C_6; this favours axial addition.

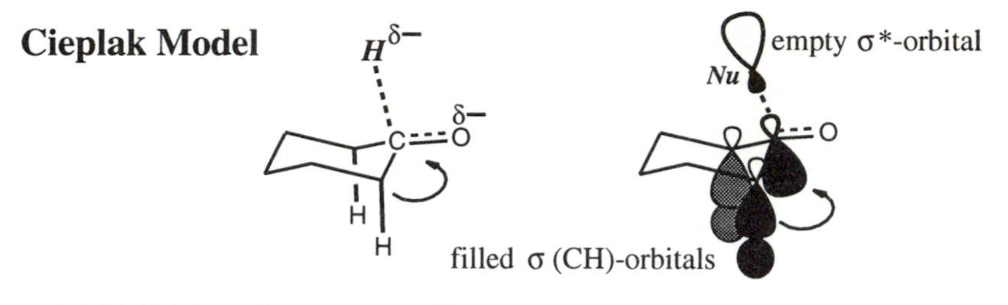

4.6 Neighbouring group effects

The stereochemistry of displacement reactions on alicyclic rings can be modified by the presence of neighbouring groups. For example, both *cis-* and *trans-*2-acetoxycyclohexyl tosylate are acetolysed (reacted in AcOH/AcO^-) to give *trans-*1,2-diacetoxycyclohexane as the sole product.

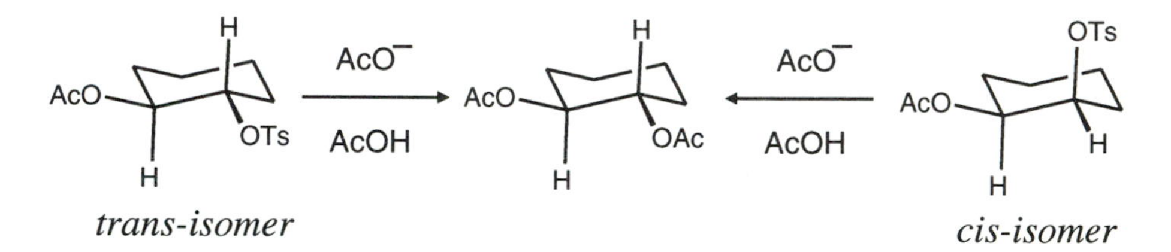

If optically active *trans-*tosylate is used, it is found that pure racemic *trans-*diacetate product is formed. Finally it is observed that the *trans* isomer reacts *ca.* 660 times faster than the *cis* (at *ca.* 100°C).

The explanation for these observations arises from the fact that two different reaction pathways are occurring. The *cis-*isomer reacts by a simple S_N2 displacement (resulting in retention of optical activity).

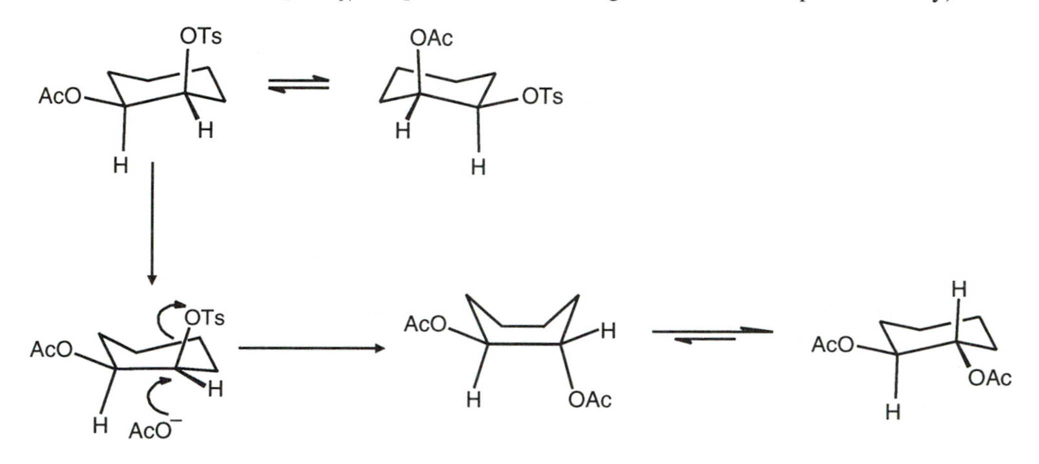

However, neighbouring group participation is possible in the *trans*-ester. The π-electrons from the acetoxycarbonyl group participate in displacement of tosylate (thereby facilitating its departure and enhancing the reaction rate) to form an intermediate which has a time-averaged plane of symmetry (hence the observed racemisation in this case) and in which the carbocation centre is efficiently stabilised by two oxygen atoms (a dioxalenium cation intermediate).

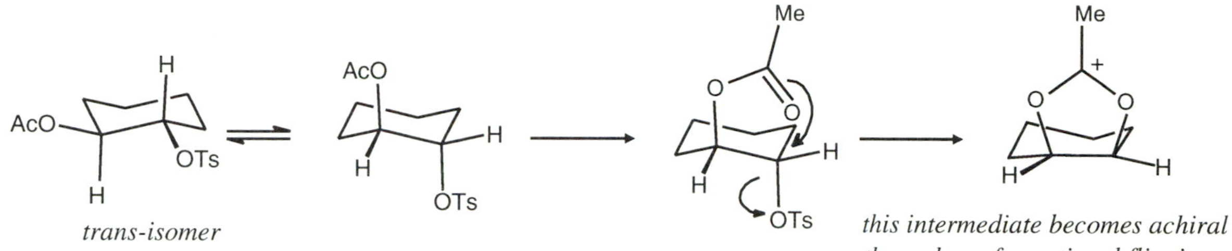

As a result of neighbouring group participation the external nucleophile must attack the intermediate from the same face as that from which tosylate departed, but this can occur at *either* the carbon atom which originally bore the tosylate group *or* at that to which the acetoxy group had originally been attached, since both sites are now equivalent. These two alternatives lead to formation of equal amounts of the enantiomeric *trans*-1,2-diacetoxycyclohexanes.

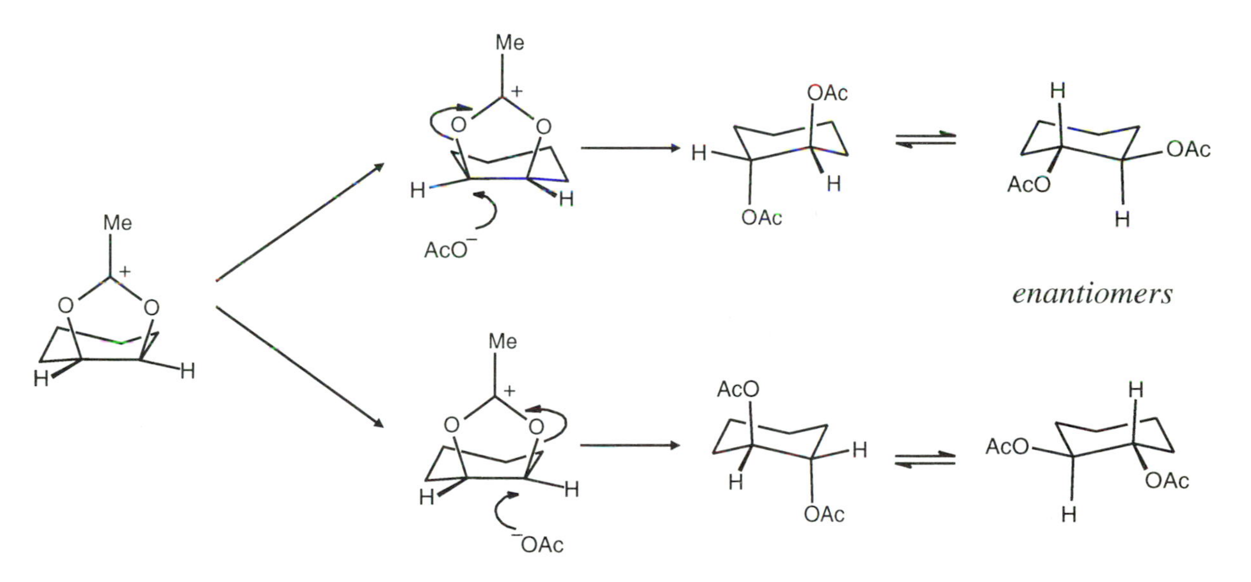

4.6.1 Hydroxylation of alkenes The above process has been exploited for the hydroxylation of alkenes in the Prévost reaction. Treatment of cyclohexene with iodine and silver acetate in acetic acid, followed by hydrolysis of the ester products affords either *cis*- or *trans*-cyclohexane-1,2-diol depending on whether wet or dry conditions respectively were used. An initially formed iodonium ion is captured by acetate to give *trans*-1-iodo-2-acetoxy-cyclohexane.

Ag(I) then induces loss of iodide which is accompanied by neighbouring group participation by the acetoxy group forming the intermediate seen previously (above). Under dry conditions this is captured by acetate ion to give the *trans*-diacetate which can subsequently be hydrolysed to the *trans*-1,2-diol.

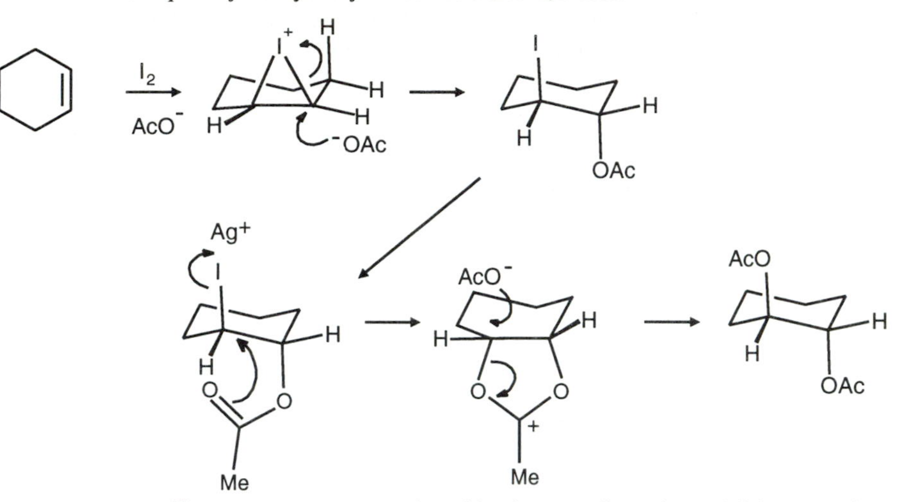

However, water attacks this intermediate by addition to the carbocation centre to give a hemi-orthoester which can fragment to *cis*-2-acetoxy-cyclohexanol, further hydrolysis of which gives *cis*-cyclohexane-1,2-diol.

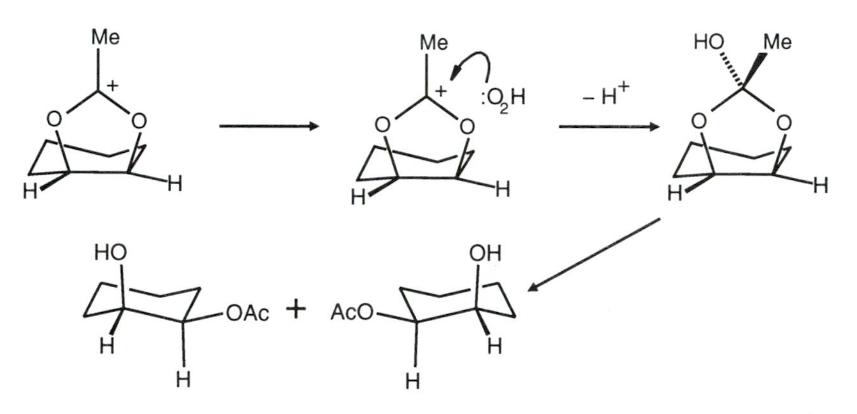

The reactions outlined in this section complement other methods for the conversion of cycloalkenes to *cis* and *trans*-1,2-diols such as epoxidation followed by hydrolysis (affording the *trans*-diol) and *cis*-hydroxylation using osmium tetroxide or aqueous potassium permanganate.

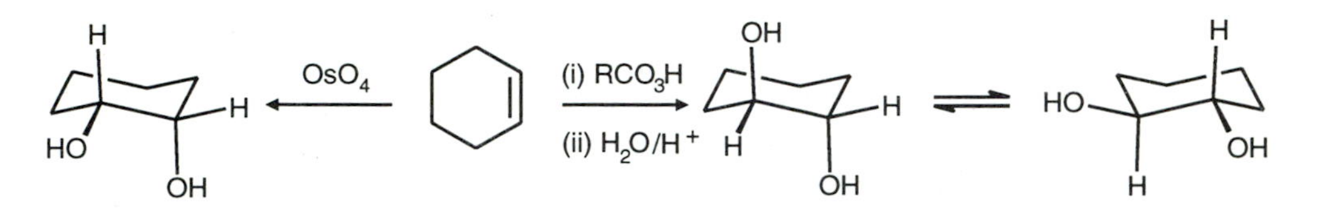

4.7 Rearrangements and transannular reactions

4.7.1 Wagner-Meerwein rearrangements The cyclohexane-1,2-diols resulting from hydroxylation of 1,2-dimethylcyclohexene can, in the presence of 20% H_2SO_4, undergo a pinacol-pinacolone rearrangement. For example, in the *cis*-diol, one Me and one HO group are *trans*-diaxial and 2,2-dimethyl-cyclohexanone is formed by methyl group migration. The *trans*-diol, in which both OH groups are axial, rearranges by a ring contraction to give methyl-(1-methylcyclopentyl)ketone (this involves migration of the ring carbon-carbon bond which is antiperiplanar to the $-OH_2^+$ leaving group).

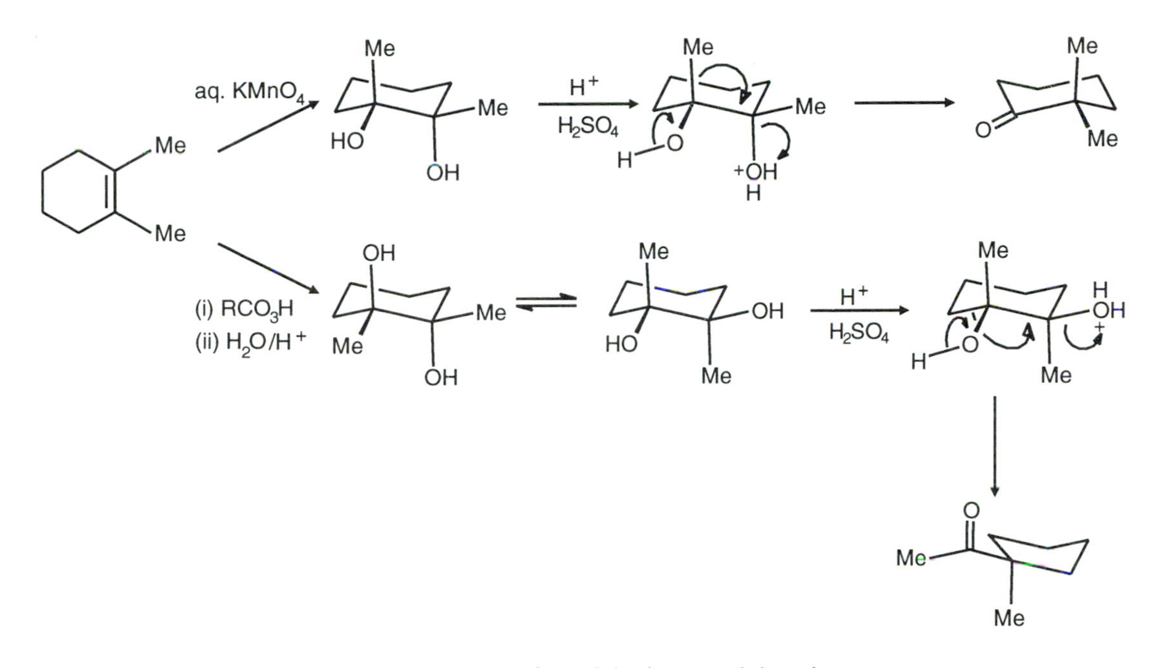

Similar rearrangements can occur when 2-hydroxycyclohexyl amines are diazotised. The use of conformationally locked amines (through the presence of a 4-*tert*-butyl group) demonstrates that rearrangement always occurs by migration of the bond *antiperiplanar* to the diazonium ion intermediate.

When both the hydroxyl and amino groups are axial, diazotisation is followed by an internal S_N2 displacement of nitrogen by the oxygen nucleophile resulting in the formation of an epoxide:

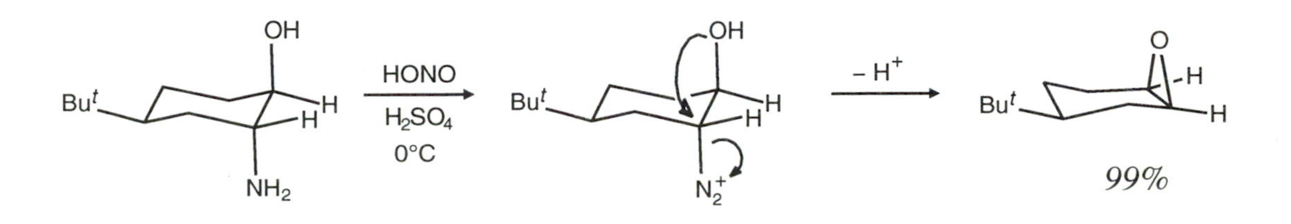

When the hydroxyl is equatorial and the amino group is axial, the substituent *trans*-coplanar with the leaving group is now a C-H bond and a 1,2-hydride shift occurs leaving an oxygen–lone-pair–stabilised carbocation which forms a carbonyl group after loss of a proton:

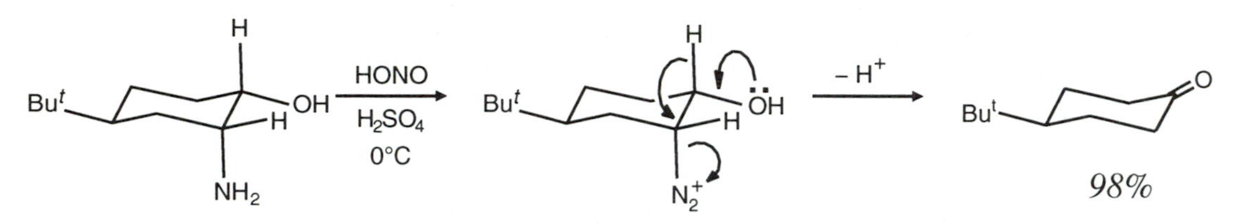

When the amino substituent is equatorial neither the O-H nor the C-H bond have an antiperiplanar relationship with the C-NH_2 bond. Instead it is a C-C bond in the ring skeleton which is now optimally placed for neighbouring group participation, resulting in ring contraction to a hydroxyl-stabilised carbocation which loses a proton to form an aldehyde:

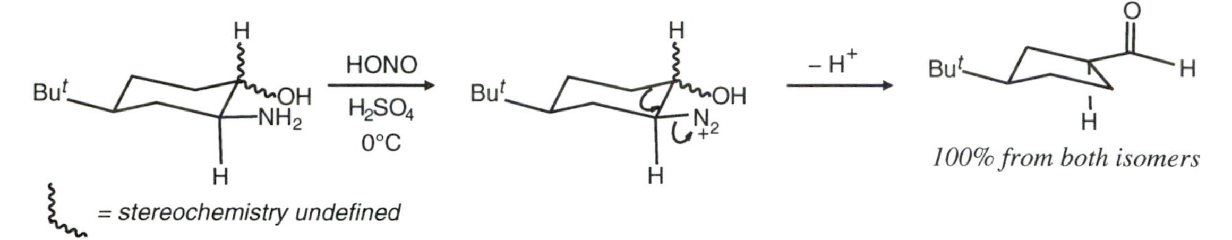

Similar stereochemical control operates in the acetolysis of decalin tosylates. When the tosylate leaving group is axial there is a methyl substituent antiperiplanar to it and a 1,2-methyl migration (a Wagner-Meerwein shift) occurs leading to the formation of a more stable tertiary cation. Elimination follows resulting in cyclohexene formation.

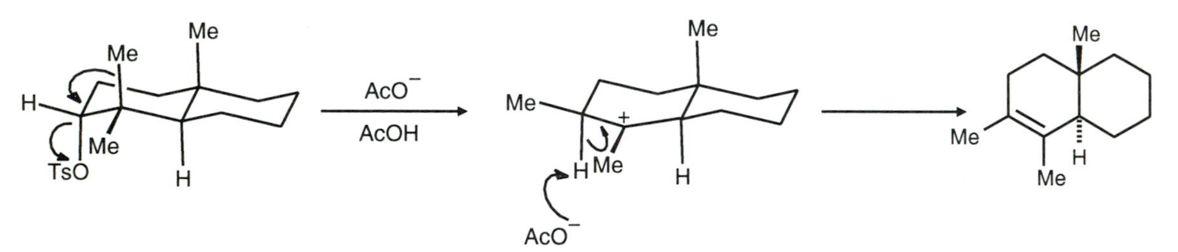

An equatorial leaving group only has ring C-C bonds antiperiplanar to it and so in this case a ring contraction is seen, the

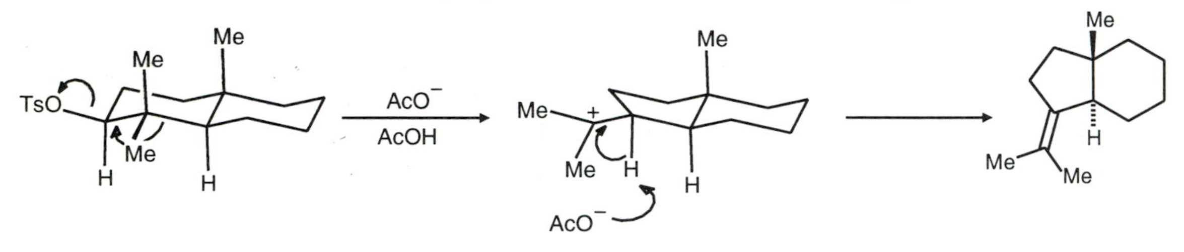

driving force once again being the rearrangement of a secondary to a tertiary carbocation.

4.7.2 Transannular rearrangements in medium rings One aspect of the strained conformations of medium rings is that opposite sides of the ring are brought into close proximity. This adds some interesting complexities to their chemistry. When carbocation intermediates are formed in medium rings, 1,3-, 1,4-, 1,5- and 1,6-hydride shifts are frequently observed. Transannular hydride transfers are particularly common during the acid-catalysed opening of medium ring epoxides. For example, rearranged products predominate in the peroxyformylation of cyclo-octene shown below. Labelling experiments confirm that both 1,3- and 1,5-hydride shifts are involved:

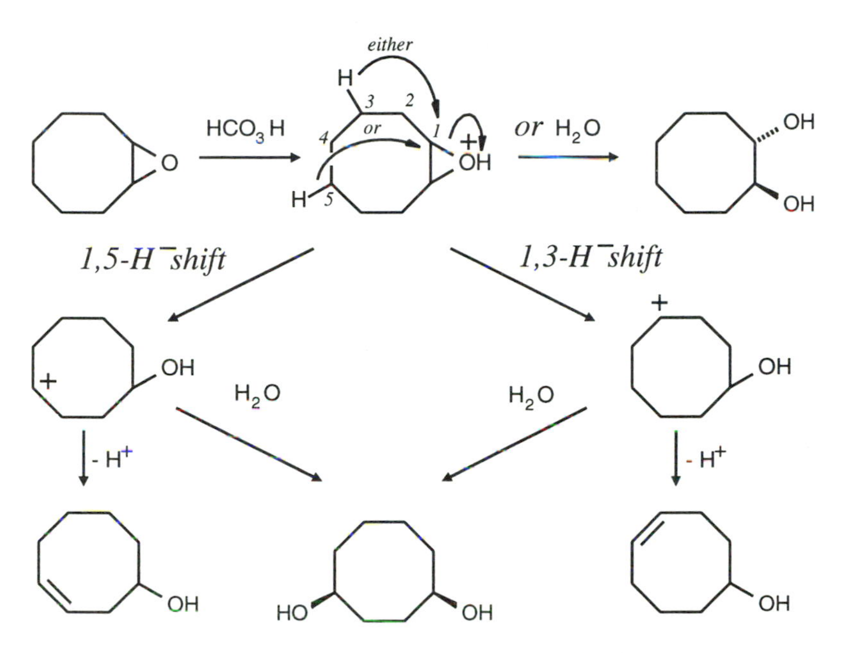

Using the deuterium labelled epoxide shown below it has been found that 1,5-hydride migration accounts for 94% of the cyclo-oct-3-en-1-ol and 61% of the 1,4-diol products.

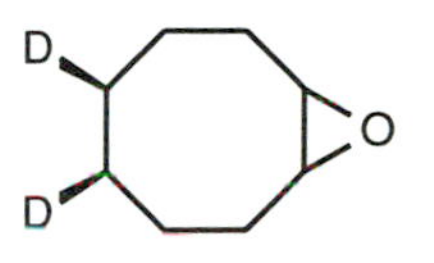

However, this reaction does not proceed by the simple mechanism suggested above since only the *trans*-1,2- and *cis*-1,4-diols are formed. This reaction cannot therefore involve free carbocations nor does the 1,5-hydride shift proceed by a concerted pathway as ring deuteriation shows only very small kinetic isotope effects [indicating that the remote C–H(D) bond is not significantly broken in the rate-determining transition state]. It has been suggested that a 1,5-hydride-bridged carbocation intermediate may be formed after rate-determining epoxide ring opening:

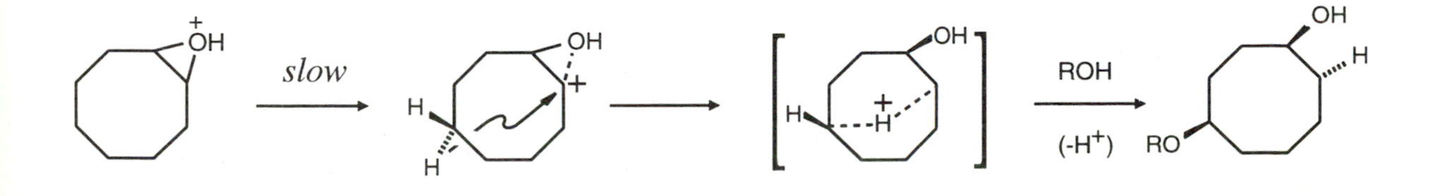

Several products are also formed from the bromination of cyclo-octene, cyclononene and cyclodecene, the two bromine substituents being found at various sites around the ring, whereas larger cycloalkenes only form the expected *trans*-1,2-dibromo adduct. The product distribution from bromination of cyclo-octene illustrates this behaviour. The reaction involves a bromonium ion intermediate which can be opened by an intramolecular transannular 1,5-hydride transfer. The resulting rearranged carbocation is finally captured by bromide ion.

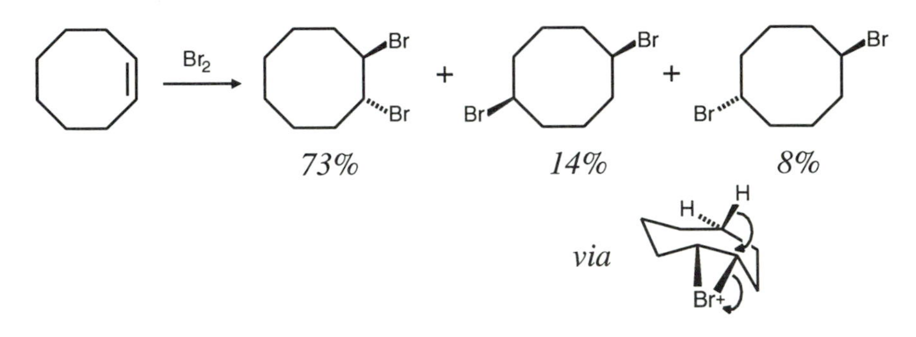

In the case of cyclodecene the only product isolated derives from internal hydride transfer:

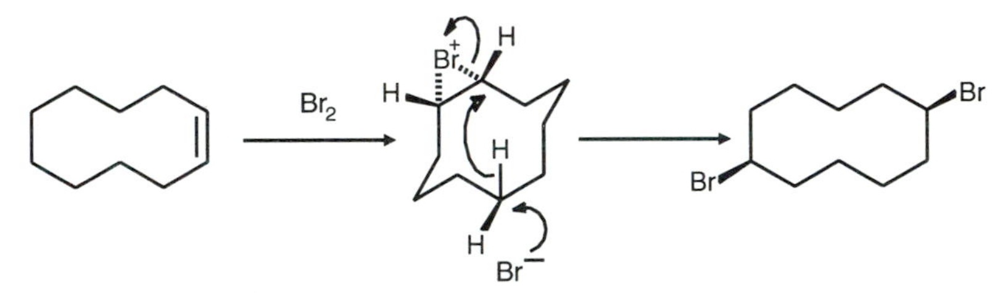

4.7.3 Transannular cyclisations Double bonds can also participate in transannular reactions. For example, acetolysis of cyclo-oct-4-enyl and cyclodec-5-enyl sulfonate esters leads to bicyclic products resulting from π-electron participation in the displacement of the sulfonate leaving group.

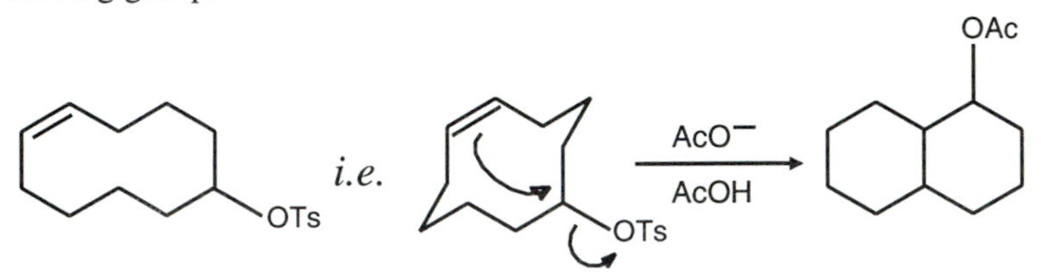

The Grob fragmentation:

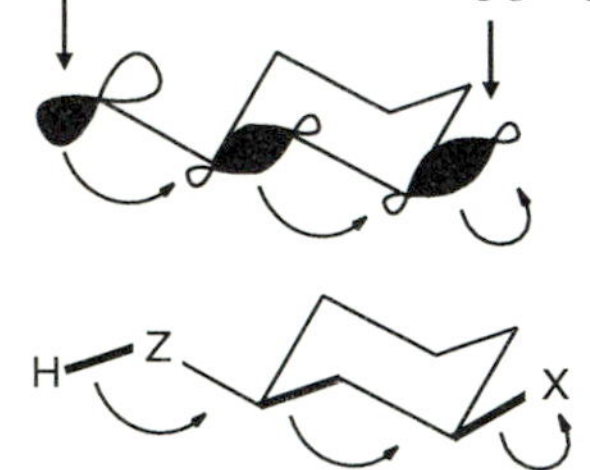

4.8 Fragmentation and other bond-cleavage reactions

4.8.1 Grob fragmentation Under certain circumstances, cyclohexyl rings incorporated into bi- and polycyclic structures can undergo carbon-carbon bond cleavage leading to the formation of a larger ring; this process is known as the Grob fragmentation. Such skeletal rearrangements are seen during the solvolysis reactions during which a carbon-carbon bond within the ring acts as an internal nucleophile helping to displace the leaving group. For this to occur there must be an antiperiplanar relationship between the bonds being broken and consequently the double bond configuration of the cycloalkene which is formed depends on the geometry of the starting material.

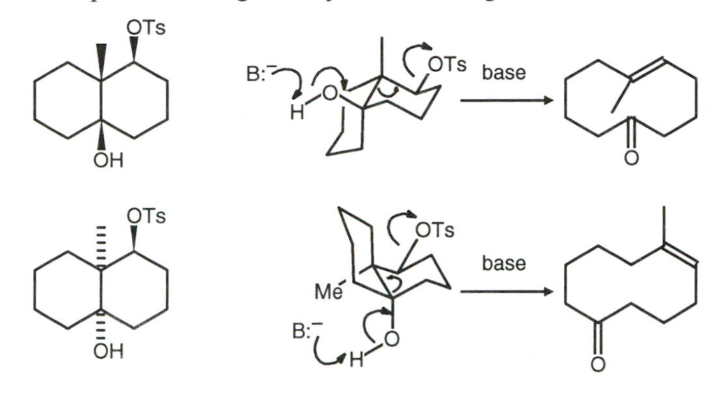

4.8.2 Oxidative cleavage of 1,2-diols Oxidative cleavage of 1,2-diols can be carried out using reagents such as HIO_4 and $Pb(OAc)_4$. Reactions of this type are thought to proceed through a cyclic intermediate and are therefore sensitive to the spatial disposition of the two hydroxyl groups involved. For example, cyclohexane *cis*-1,2-diol is cleaved by lead tetra-acetate about 22 times faster than its *trans*-isomer despite the fact that the dihedral angle between the C-OH bonds is the same in both isomers. The rate difference arises because the *cis*-isomer achieves an ideal intermediate geometry by flipping into a boat conformation - the *trans*-isomer cannot do this.

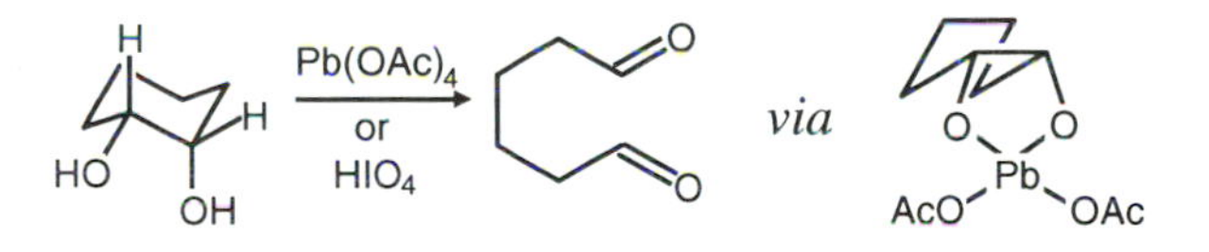

However, when the two OH groups are locked *trans*-diaxial the cyclic intermediate cannot form and cleavage does not occur with periodate - though reaction still occurs with lead tetra-acetate, probably by a different (non-cyclic) mechanism.

Oxidatve cleavage of 1,2-diols by HIO_4 or $Pb(OAc)_4$ proceeds *via* a cyclic intermediate, *e.g.*

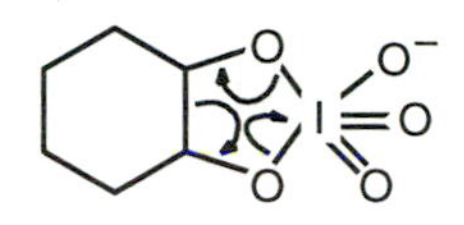

When this geometry becomes impossible, as in the *trans*-decalindiol shown below, reaction does not occur.

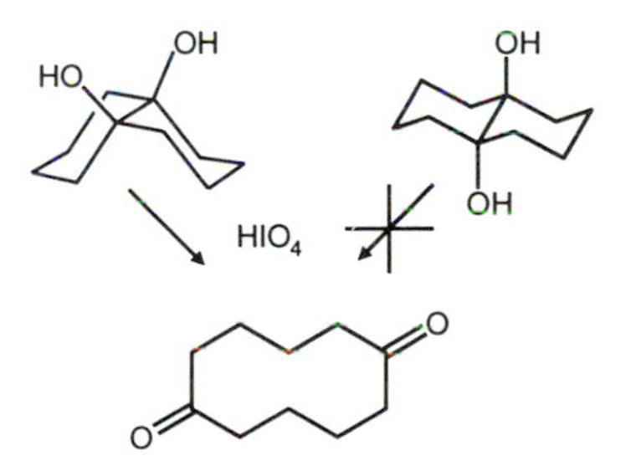

4.9 Ring expansion and contraction processes

4.9.1 The Favorskii reaction α-Bromoketones undergo base-catalysed ring contraction. For example, 2-bromocyclohexanone forms an ester of cyclopentane carboxylic acid and 2-bromocyclobutanone is converted into the cyclopropane ester. Isotopic labelling experiments show that the normal reaction pathway involves a bicyclo[n.1.0]ketone intermediate:

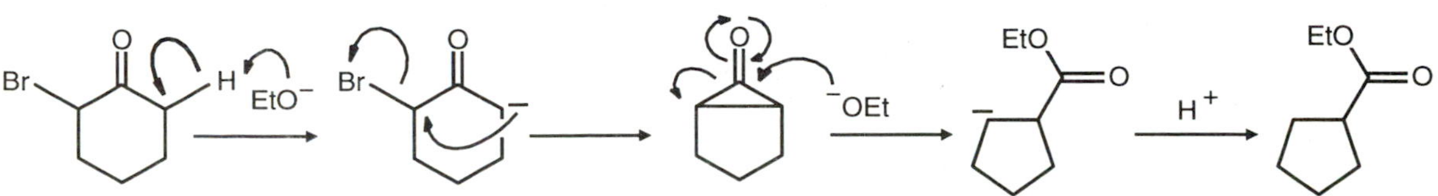

However, an alternative mechanism operates for the cyclobutanone because of the unfavourability of enolisation in the small ring. Similar behaviour is observed in other non-enolisable structures.

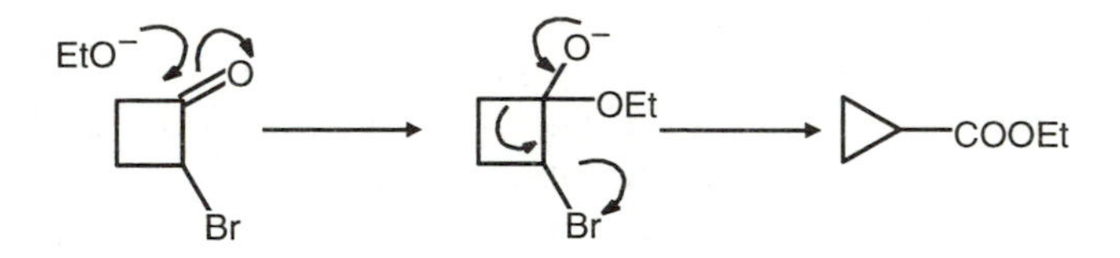

4.9.2 The Wolff rearrangement α-Diazoketones undergo photochemically induced ring contraction, probably *via* a carbene intermediate, to give a ketene. This is rapidly hydrated to form the corresponding carboxylic acid or, in the presence of amines or alcohols, forms amides or esters respectively. This process is well established for $C_6 \rightarrow C_5$ and $C_5 \rightarrow C_4$ ring contractions.

Ring contraction of a carbene intermediate is a key step in the Wolff rearrangement:

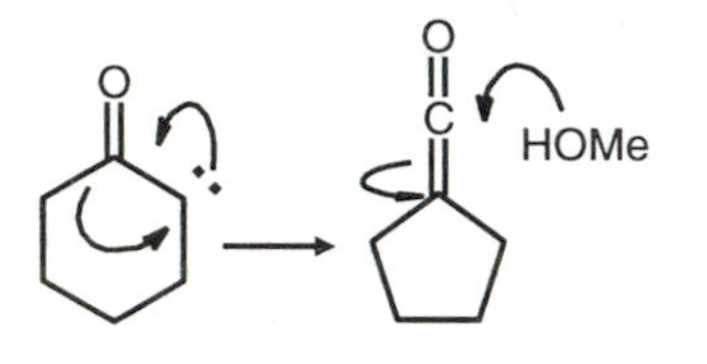

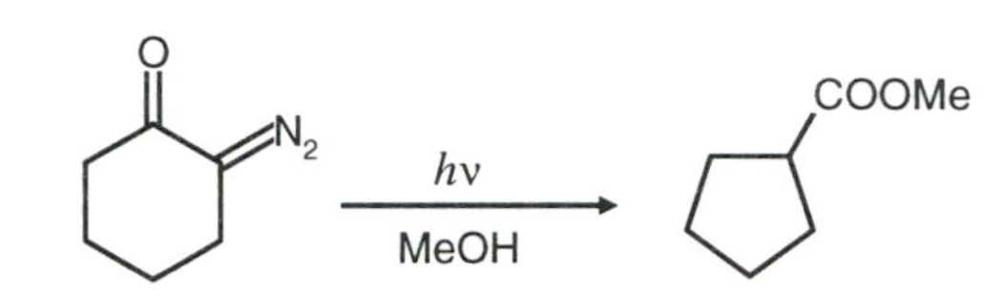

4.9.3 The Demjanov ring expansion Cycloheptanone is the major product of the reaction of cyclohexanone with diazomethane (*route a*), accompanied by some (10%) epoxide by-product (*route b*). However this process is of limited value since it requires the starting ketone to be more reactive towards addition of diazomethane than the product, a situation which is rare; for example, cyclopentanone gives a mixture of both C_6 and C_7 ketone product and yields from larger ring ketones tend to be low.

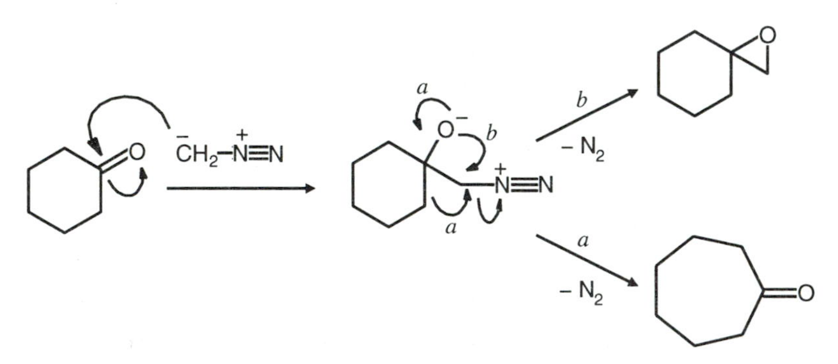

Lewis acids improve reaction efficiency; co-ordination reduces the nucleophilicity of the alkoxide intermediate thereby impeding exocyclic epoxide formation. This diazo-intermediate can also be accessed by deamination (using RONO) of a 2-aminoalcohol obtained by hydride reduction of a cyanohydrin (the Demjanov-Tiffeneau ring expansion), an approach which gives improved yields for ketones in the range C_4 to C_8.

5. Polycyclic systems

5.1 Bridged and caged structures

5.1.1 Nomenclature Bi- and polycyclic hydrocarbons are named according to the way in which the component rings are fused to each other. It is first important to recognise the bridgehead atoms (those atoms which are common to different rings. Then:

(i) the number of rings that share atoms defines the "...cyclo" prefix, *i.e.* 2 rings = bicyclo, 5 rings = pentacyclo, etc.;

(ii) this is followed by a series of numbers (in square brackets and separated by full stops) which denote the number of atoms in each bridge linking the bridgehead atoms starting with the longest chain;

(iii) and finally there follows the hydrocarbon name reflecting the total number of carbons atoms in the structure (together with details of any substituents following the usual rules).

Such structures are numbered starting at a bridgehead carbon (= C_1) and moving along each connecting chain to the other bridgehead in sequence, working from the longest down to the shortest chain.

Norbornane is called:

bicyclo[2.2.1]heptane

and is numbered:

5.1.2 Bicycloheptanes One of the best studied bicyclic systems is bicyclo[2.2.1]heptane also known as norbornane. Derivatives of this hydrocarbon have a number of interesting properties partly arising from strain present in the structure, and provide a useful framework for testing aspects of reactivity and mechanistic pathways, particularly because of the rigid nature of the skeleton which facilitates well-defined stereochemical control. Perhaps one of the more remarkable features of the reactions of 2-norbornyl derivatives is that a number of skeletal rearrangements can occur during reactions in which the 2-norbornyl cation is an intermediate, but these simply regenerate another 2-norbornyl cation (but one in which the carbocation centre has moved to another location).

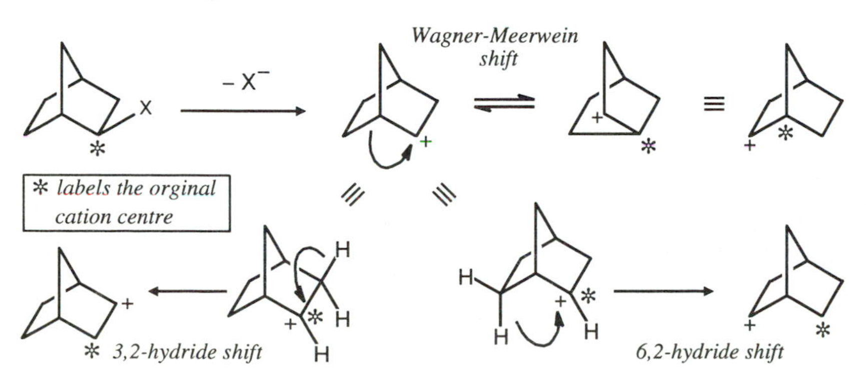

The nature of the 2-norbornyl cation has been a matter of considerable controversy. It behaves as a non-classical cation (*i.e.* a structure in which there is a 2-electron 3-centre bond) or a pair of classical but rapidly interconverting cations:

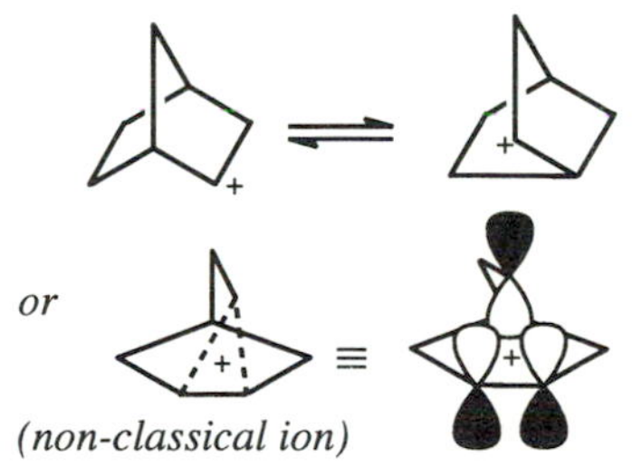

At room temperature, ^1H and ^{13}C nmr spectra of solutions of the 2-norbornyl cation show a single resonance because of fast 3,2- and 6,2- hydride shifts coupled with site scrambling by a non-classical (or the rapid equilibration of a pair of classical) cations.

5.2 Bredt's rule

In bicyclic and polycyclic structures it is sometimes not possible to locate a double bond at a bridgehead. This is a particular problem when the rings are small, as is the case in norborn-1-ene, since it is not possible to sustain an approximately planar geometry about each of the alkenyl carbon atoms as required for satisfactory p-orbital overlap to be maintained.

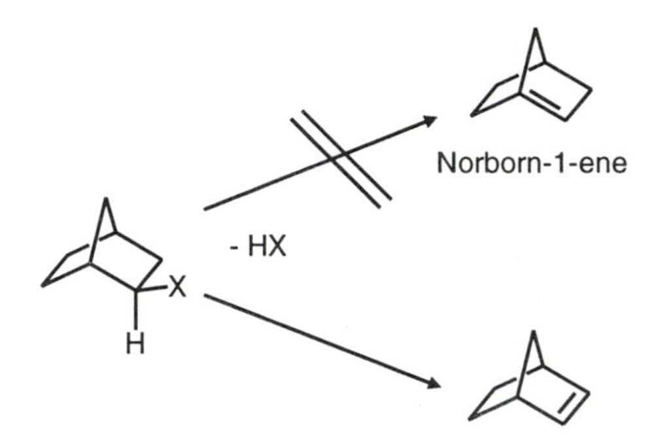

Bredt's rule states that, in a small bridged ring system, it is not possible, for reasons of excessive strain, to have a double bond at a bridgehead position.

This is the basis of *Bredt's rule* which implies that an elimination process in a bridged structure should always avoid the formation of an alkene at a bridgehead. It is therefore useful to try to determine the minimum ring sizes which will accommodate a bridgehead double bond. In a series of bicyclononenes it has been found that the [3.3.1]non-1-ene and the [4.2.1]non-1(8)-ene are relatively stable whereas the [3.3.2]non-1-ene isomer is as yet unknown. As a rule of thumb it is useful to look at the largest ring component in the structure and compare it with the corresponding simple cycloalkene. For example, the bicyclo[3.3.2]non-1-ene shown below, which dimerises as soon as it is formed, can be viewed as a derivative of *trans*-cycloheptene which is also only known as a short-lived intermediate. On the other hand the other two isomers shown can be viewed as *trans*-cyclo-octene derivatives and indeed have strain energies of a similar order of magnitude to that compound. Bicyclo[3.3.1]non-1(2)-ene has been prepared optically active and, in contrast with *trans*-cyclo-octene, cannot racemise because the single methylene bridge prevents rotation into the *trans*-chair conformation.

The limits of Bredt's rule

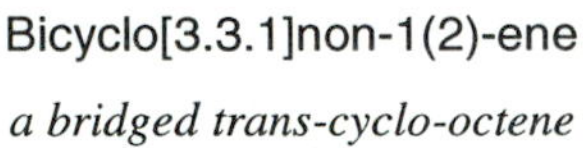

Bicyclo[3.3.1]non-1(2)-ene	Bicyclo[4.2.1]non-1(8)-ene	Bicyclo[3.2.2]non-1-ene
a bridged trans-cyclo-octene	*a bridged trans-cyclo-octene*	*a bridged trans-cycloheptene*
known	***known***	***unknown***

Whilst adamantene has not been prepared 4-adamantyl homoadamantane is very stable, despite being a *trans*-cycloheptene derivative:

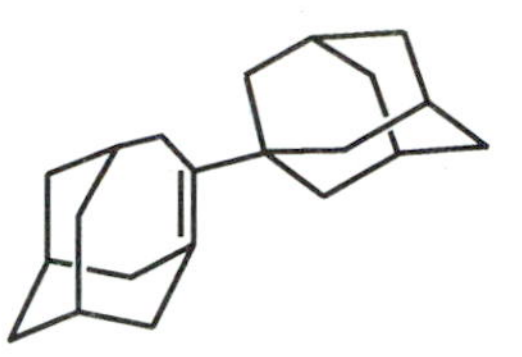

The steric bulk of the adamantyl substituent on the double bond prevents dimerisation (*via* a thermal [2+2]-cycloaddition process) which is the normal decomposition pathway for such unstable alkenes.

Small bicycloalkenes containing a *trans*-cyclohexene moiety are very unstable and have only been detected by trapping.

5.3 Isomerism in bicyclanes

Norbornane (bicyclo[2.2.1]heptane) can be viewed as a bridged boat cyclohexane. In this structure the two bridgehead C-H bonds are pointing outwards relative to the carbocyclic skeleton; it would therefore be possible to have other geometric arrangements at these centres. Clearly no alternatives are possible in this very strained structure, but if the bridging chains were longer, bridgehead isomerism should be possible as is shown below.

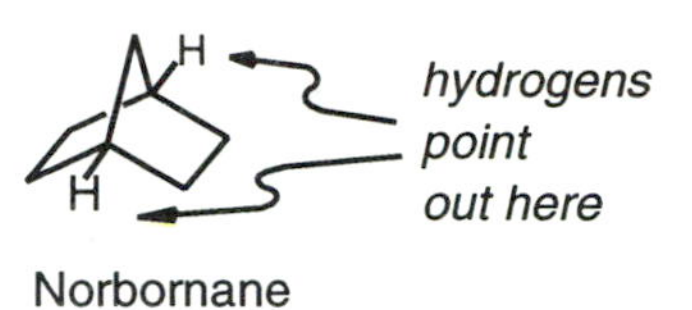

Norbornane

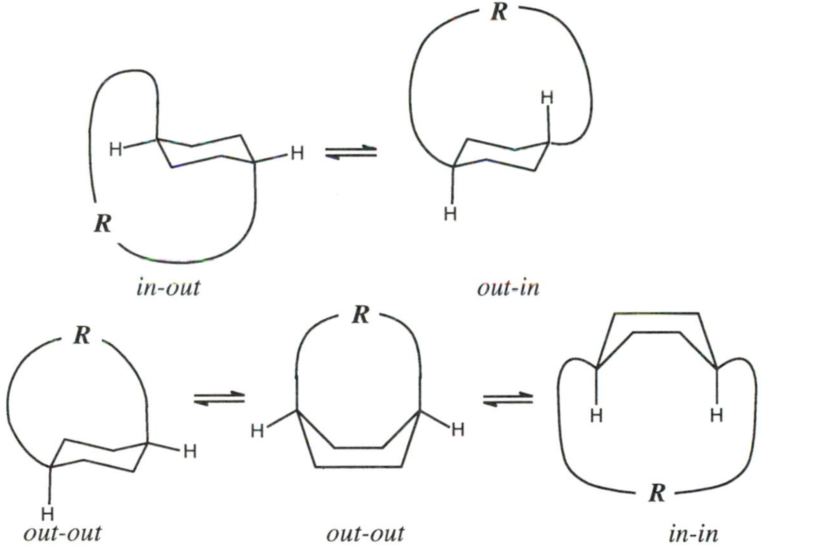

Some bicyclanes:

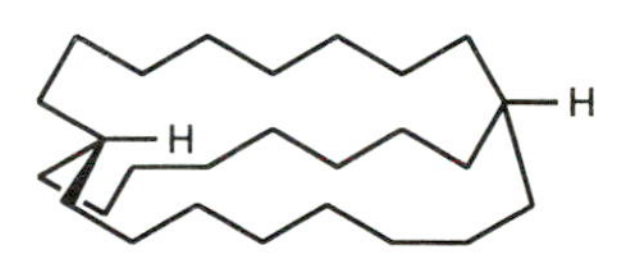

in-out-Bicyclo[8.8.8]hexacosane

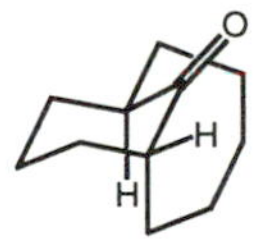

trans-Bicyclo[5.3.1]undecan-11-one

The first such isomers synthesised were "in-in" and "in-out" bicyclo[8.8.8]hexacosanes and the smallest "in-out" isomer so far prepared is *trans*-bicyclo[5.3.1]undecan-11-one which rather surprisingly does not isomerise to the more stable and known "out-out" isomer, a reflection perhaps of the fact that this process would require enolisation *via* an "anti-Bredt" intermediate.

5.4 Propellanes

These are compounds in which two carbon atoms, directly connected, are also linked by three other bridges. In norbornane the bridgehead bonds must point outwards whereas in the corresponding propellane they must point inwards to form the bridgehead carbon-carbon bond. The result is a [2.2.1]propellane.

When such structures are viewed along the bridgehead-bridgehead bond they look like propellers, hence the name "propellanes". In such structures the "saturated" (*i.e.* tetravalent) bridgehead carbon atoms must have all their bonds on one face of a plane leading to very low stability.

Propellanes can be prepared by the addition of carbenes across bicycloalkenes or by elimination of bromine from bicycloalkanes bearing bromine substituents at the bridgeheads.

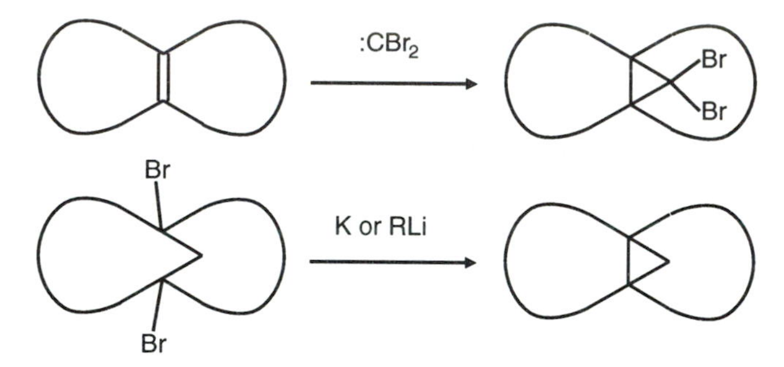

[4.2.2]- [3.3.2]- and [3.2.1]-propellane are all relatively thermally stable (up to 160°C) but the [4.2.1]-isomer is less so. All are sensitive to electrophilic cleavage of the central C-C bond, indeed this bond is cleaved in preference to addition to an alkene substituent within the same molecule:

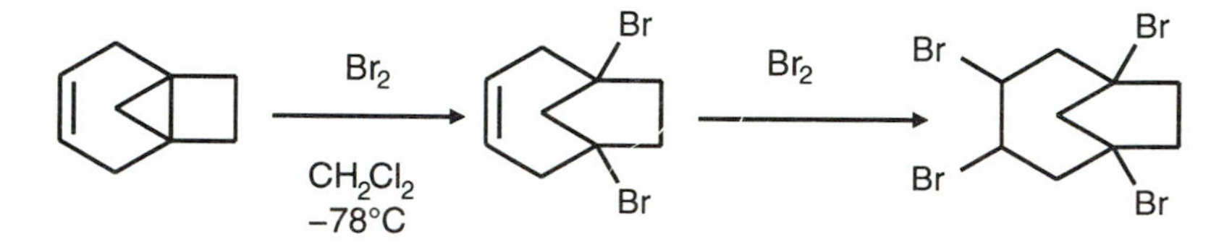

[1.1.1]Propellane has a very long central C-C bond (*ca.* 160 pm) reflecting the strain present in this structure.

The lowest member of the series, [1.1.1]propellane, is in fact more stable than its [2.1.1]- and [2.2.1]-homologues. The relative stability of [1.1.1]propellane (its infrared spectrum has been recorded at room temperature and its structure has been determined by electron diffraction) reflects the fact that the intermediate arising from central C–C bond cleavage is also highly strained and therefore does not offer a convenient pathway for lowering the energy of the system.

5.5 Cubane and other "Platonic solids"

The Platonic solids, polyhedra of high symmetry, represent fascinating targets for synthesis and study. These structures include the tetrahedron, the cube, the octahedron, the dodecahedron, and the icosahedron.

Cubane is perhaps one of the most aesthetically pleasing of all organic solids - a Platonic solid (*i.e.* a polyhedron of high symmetry). It was first synthesised in 1964 by Eaton and Cole by route which exploited a photochemically induced {2+2]-cycloaddition and a Favorskii ring contraction as two of the key steps. A simpler route, involving trapping of cyclobutadiene was subsequently devised. The key steps of both are summarised below:

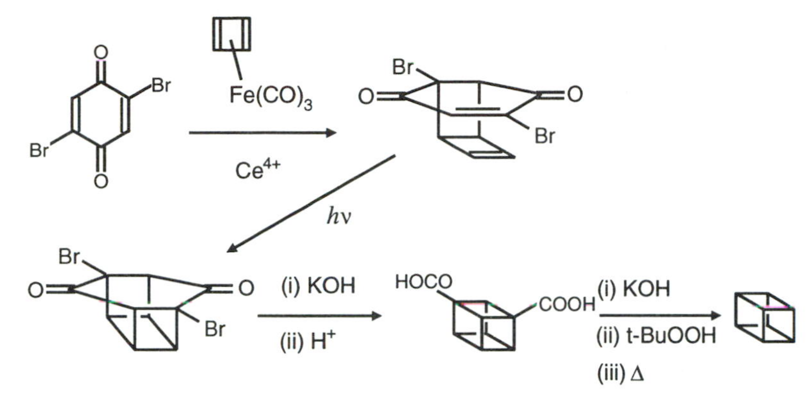

An X-ray structural view of cubane is shown below:

An indication of the electronic structure of cubane comes from nmr spectroscopy, the coupling $J(^1H-^{13}C)$ having a value of 160 Hz indicative of significant s-character in the C-H bonds as a result of the need for considerable p-character in the C-C σ-bonds required to accommodate the 90° bond angles.

Dodecahedrane itself was first prepared in 1982:

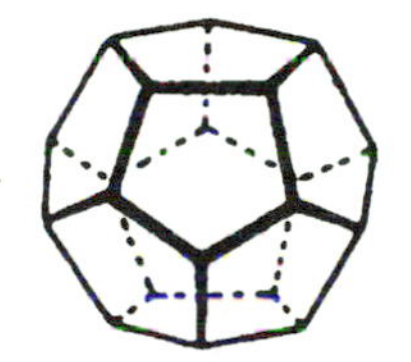

Other Platonic solids, *e.g.* dodecahedrane and tetrahedrane derivatives have also been prepared, the latter being most readily obtained as its tetra-*tert*-butyl derivative:

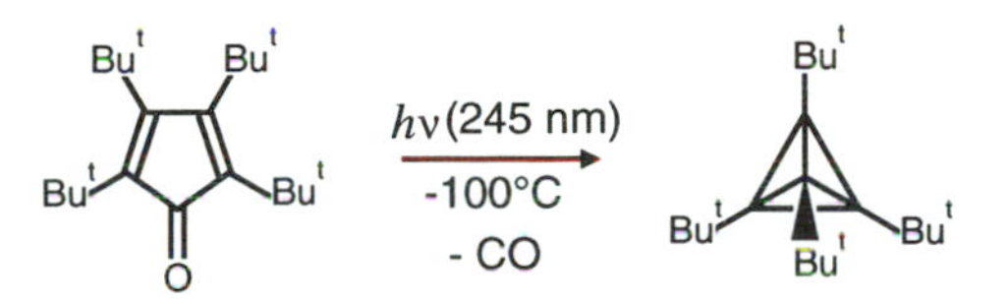

An X-ray structural study reveals "normal" C-C bond lengths and a C-C-C bond angle of *ca.* 108°. Shown below is the carbon skeleton of 1,16–dimethyl-dodecahedrane

Despite considerable strain (*ca.* 544-628 kJ mol^{-1}), this derivative of tetrahedrane melts at 135°C without decomposition. The preparation of less substituted structures has proved much more difficult and it is thought that the tetra-*tert*-butyl derivative is stabilised by a "corset effect" since the tetrahedral geometry keeps the bulky *tert*-butyl groups as far away from each other as possible thereby minimising steric interference (an X-ray structural view of the carbon skeleton is shown below).

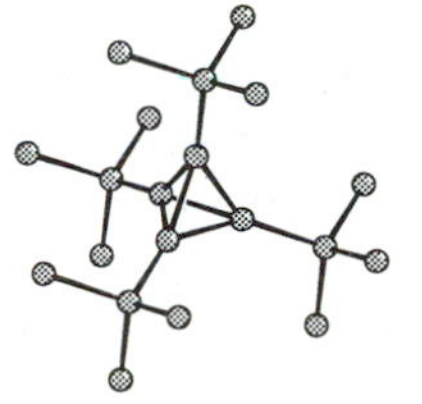

5.6 Catenanes

The other great challenge for alicyclic chemistry is the construction of chains of interlocked rings. This was first achieved in *ca.* 0.1% yield using a statistical approach involving an acyloin reaction by Wasserman in 1960. In this experiment the acyloin reaction shown below was carried out in a deuterium-labelled hydrocarbon solvent. When the acyloin product was investigated, it was found to contain a very small amount of deuterium label which had arisen from the threading of the diester through the large hydrocarbon macrocycle and subsequent ring closure.

Under normal conditions catenanes (in which two macrocycles are interlocked as the first pair of links of a chain) are formed in very low yield.

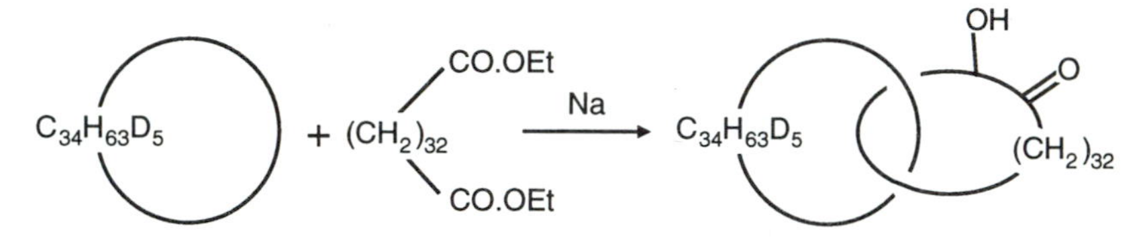

More recently intermolecular interactions have been exploited to template such syntheses and catenanes can now be made in yields comparable with those of more traditional reactions (*e.g.* as high as 70%). Indeed Stoddart, a pioneer in the field, has now successfully synthesised an Olympic ring symbol (*i.e.* 5 interlocking rings). An example of this approach to catenanes is illustrated below:

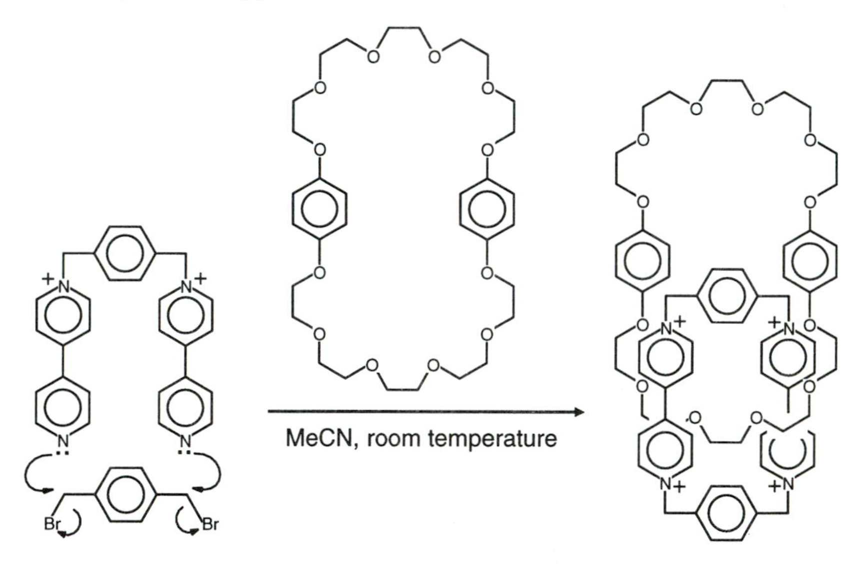

When the pyridine ring alkylates, the resulting trication threads through the crown ether in such a manner that the final alkylation step of catenane formation becomes very favourable. This final step is favoured by a series of aromatic associative interactions and hydrogen bonding which help to pre-organise the structure of the intermediate.

Index